I0482660

Disclaimer

The publisher of this book is by no way associated with the National Institute of Standards and Technology (NIST). The NIST did not publish this book. It was published by 50 page publications under the public domain license.

50 Page Publications.

Book Title: A Literature Review of the Effects of Fire Smoke on Electrical Equipment

Book Author: Richard D. Peacock; Thomas G. Cleary; Paul A. Reneke; Daniel Murphy;

Book Abstract: A review is presented of the state of the art of smoke production measurement, prediction of smoke impact as part of computer-based fire modeling, and measurement and prediction of the impact of smoke through deposition of soot on and corrosion of electrical equipment. The literature review on smoke corrosivity testing and damage due to smoke deposition emphasizes (despite extensive research on smoke corrositity) the lack of validated and widely applicable prescriptive or performance based methods to assure electrical equipment survivability given exposure to fire smoke. Circuit bridging via current leakage through deposited smoke was identified as an important mechanism of electronic and electrical equipment failure during NPP fires. In the near term, assessment of potential damage can reasonably be based on the airborne smoke exposure concentration and, perhaps, the exposure duration. Hence, models that can predict the airborne smoke concentration would be sufficient to suit short-term analysis needs. In the longer term, it would be desirable to develop models that could estimate the deposition behavior of smoke, as well and specifically correlate the combination of deposited and airborne smoke to component damage.

Citation: NIST TN - 1769

Keywords: cables; heat release rate; fire tests; smoke; standards; test methods; wires

NUREG/CR-7123

A Literature Review of the Effects of Smoke from a Fire on Electrical Equipment

Office of Nuclear Regulatory Research

NUREG/CR-7123

A Literature Review of the Effects of Smoke from a Fire on Electrical Equipment

Manuscript Completed: November 2011
Date Published: July 2012

Prepared by:
Richard D. Peacock, Thomas G. Cleary, Paul A. Reneke, Daniel C. Murphy
Engineering Laboratory

National Institute of Standards and Technology
Gaithersburg, Maryland 20899

David Stroup, NRC Project Manager

Prepared for:

**Division of Risk Analysis
Office of Nuclear Regulatory Research
U.S. Nuclear Regulatory Commission
Washington, DC 20555-0001**

Office of Nuclear Regulatory Research

ABSTRACT

A review is presented of the state of the art of smoke production measurement, prediction of smoke impact as part of computer-based fire modeling, and measurement and prediction of the impact of smoke through deposition of soot on and corrosion of electrical equipment. The literature review on smoke corrosivity testing and damage due to smoke deposition emphasizes (despite extensive research on smoke corrosivity) the lack of validated and widely applicable prescriptive or performance based methods to assure electrical equipment survivability given exposure to smoke from a fire. Circuit bridging via current leakage through deposited smoke was identified as a potentially important mechanism of electronic and electrical equipment failure during nuclear power plant fires.

In the near term, assessment of potential damage can reasonably be based on the airborne smoke exposure concentration and the exposure duration. Hence, models that can predict the airborne smoke concentration would be sufficient to provide upper limit estimates of potential damage. In the longer term, it would be desirable to develop models that could estimate the deposition behavior of smoke, and specifically correlate the combination of deposited and airborne smoke to component damage.

CONTENTS

LIST OF FIGURES

LIST OF TABLES

ACKNOWLEDGEMENTS

The work described in this report was supported by the Office of Nuclear Regulatory Research (RES) of the US Nuclear Regulatory Commission (USNRC). The project was directed by David Stroup of the Fire Research Branch (FRB), Division of Risk Analysis (DRA) in RES.

ABBREVIATIONS AND ACRONYMS

C	Carbon
CFAST	Consolidated Model of Fire Growth and Smoke Transport
CMOS	Complimentary Metal-oxide Semiconductor
CNET	Centre National d'Études des Télécommunications
CO_2	Carbon Dioxide
DB-9	D-subminiature, 9-pin
DB-25	D-subminiature, 25-pin
DIP	Dual Inline Package
DTC	Digital Trip Computer
EDSC	Experimental Digital Safety Channel
EQ	Environmental Qualification
EPRI	Electric Power Research Institute
EPROM	Erasable Programmable Read-only Memory
EVA	Ethylene-vinyl Acetate
FDDI	Fiber Distributed Data Interchange
FDS	Fire Dynamics Simulator
FHA	Fire Hazards Analysis
FMRC	Factory Mutual Research Corporation
FOM	Fiber-optic Module
FR-6400	Low Density Polyethylene with Fire Retardant
FR-EVA	Fire-retardant Ethylene-vinyl Acetate
HOSTP	Host Processor
HCl	Hydrogen Chloride
HCLV	High-current Low-voltage
HDD	Hard Disk Drive
HF	Hydrogen Fluoride
HFLPF	High-frequency Low-pass Filter
HFTL	High-frequency Transmission
HRR	Heat Release Rate
HSD	High-speed Digital
HVLC	High-voltage Low-current
IEC	International Electrochemical Commission
LAN	Local Area Network
LDPE	Low Density Polyethylene
MCC	Motor Control Center
MLR	Mass Loss Rate
NEC	National Electrical Code
NFPA	National Fire Protection Association
NIST	National Institute of Standards and Technology
NO_2	Nitrogen Dioxide
NPP	Nuclear Power Plant
NRC	Nuclear Regulatory Commission
OSU	Ohio State University
PCB	Printed Circuit Board

PE	Polyethylene
PE165	Polyethylene with Fire Retardant
PFPC	Polyolefins Fire Performance Council
PLLC	Plastic Leadless Chip Carrier
PRA	Probabilistic Risk Assessment
PRS/MUX	Process Multiplexer Unit
PTH	Pin Through Hole
PVC	Poly-vinyl Chloride
QCM	Quartz Crystal Microbalances
RES	NRC Office of Nuclear Regulatory Research
RH	Relative Humidity
RI/PB	Risk-informed, Performance-based
RJ-45	Network Modular Connector
SMT	Surface Mount Technology
SNL	Sandia National Laboratories
SO$_2$	Sulfur Dioxide
SRAM	Statis Random Access Memory
TTL	Transtitor-to-transistor Logic
UL	Underwriters Laboratories

1   BACKGROUND

1.1   Introduction and Purpose

The Browns Ferry nuclear power plant (NPP) fire in 1975 demonstrated that instrument, control, and power cables are susceptible to fire damage [1-3]. At Browns Ferry, over 1,600 cables were damaged by the fire and caused short circuits between energized conductors. In addition to the cable damage, the fire deposited soot throughout the Unit 1 reactor building and in small areas of the Unit 2 reactor building. Examination of all surfaces of piping, conduit, and other equipment showed limited evidence of chlorine-induced corrosion, requiring replacement of affected material and an accelerated inspection program for some stainless steel within the building [4]. In addition to direct damage due to elevated temperature and heat from fires, reports also indicate non-thermal damage to equipment that was exposed to smoke and combustion gases from the fire environment (see, for example, Refs [23-5]). Although limited information is evident specifically related to nuclear power plants [4,6,7], damage to electrical systems in other industries from fire has been extensive [5,8], with some resulting in losses in the hundreds of millions of dollars [9].

Some non-thermal effects occur over long time periods and thus would largely impact clean-up and restoration after a fire incident. For example, corrosion several days after a one hour fire exposure can be several times that initially observed after the exposure [10]. However, shorter-term effects have also been documented in major electronics fires. In the Hinsdale Illinois telecommunications central office fire, smoke-induced failures were noted within six hours [5].

Over the past decade, there has been a considerable movement in the nuclear power industry to transition from prescriptive rules and practices toward the use of risk information to supplement decision-making. One element crucial in supporting the use of risk-informed applications is the availability of tools to evaluate the likelihood and consequences of fire scenarios. Risk-informed, performance-based (RI/PB) fire protection often relies on fire modeling to determine the consequences of fires. Estimating target damage is a key part of any fire modeling analysis. Methodologies are available to obtain reasonable quantitative predictions of thermal damage. However, the current state of knowledge does not support similar detailed quantitative prediction of smoke damage.

Based on previous Nuclear Regulatory Commission (NRC) testing [11], four modes of failure due to smoke damage have been identified. Of these four failure modes, only one, circuit bridging, has been found to be potentially risk significant. Current fire models and data are insufficient at this time to directly assess the risk contribution of circuit bridging faults. Screening or bounding assessments can be made, but given current knowledge, would be dependent on the application of expert judgment and would have considerable uncertainty. For example NUREG/CR-6850, "Fire PRA Methodology for Nuclear Power Facilities" [11], includes the following recommendations related to smoke damage of electrical equipment:

- "If the fire scenario involves an electrical panel, it may be prudent to assume the smoke-induced failure of all digital or integrated circuit components within the originating fire panel regardless of the assumed fire size, intensity, or duration.

- In the event of a fire involving high-energy electrical components (e.g., MCC, breaker, switchgear, etc.), it may be prudent to assume the smoke-induced failure of components in adjoining panels or cubicles, especially if those cubicles or panels are connected by features like bus ducts or a common ventilation system."

Such broad assumed failures have the potential to over- or under-estimate the hazard depending on the size and spread of fire and the susceptibility of the equipment to fire.

In order to assess the potential for damage due to smoke exposure, relationships between smoke exposure and the failure of real electronic components need to be established. A smoke damage routine developed for a fire model could then assess the near term (during and soon after exposure) damage potential of electronics and electrical components design fire exposures.

1.2 Organization of the Report

The overall objective of this research program is to develop a better understanding of how fire-induced smoke production and transport can affect electronic equipment that may be used in an NPP. This report presents a review of the measurement of the impact of smoke through deposition of soot on and corrosion of electrical equipment, smoke production measurement, and prediction of smoke impact as part of computer-based fire modeling.

Chapter 2 presents previous testing of the effects of fire-generated smoke on electronic equipment. Studies focused on NPPs and on telecommunication equipment are reviewed. A general overview of corrosion of materials exposed to smoke and of existing damage criteria that have been applied are included.

Chapter 3 discusses bench- and large-scale testing and test methods that have been used to quantify the production of smoke and its components from a fire.

Chapter 4 provides an assessment of the state of the art in modeling the generation, transport, and deposition of smoke and fire gases in current compartment fire models.

Chapter 5 identifies future research needs to improve the prediction of smoke transport and its potential impact on NPP equipment.

2 EFFECTS OF SMOKE ON ELECTRONIC EQUIPMENT

In the United States, flammability requirements for electrical equipment are most commonly specified in the National Electrical Code (NEC) [12]. Specific testing requirements are then defined by testing laboratory standards (e.g., Underwriters Laboratories, Canadian Standards Association, etc.). Such standards are essentially voluntary until required by a building code, regulatory agency, or by the building owner as part of the bid process for construction of a facility. Such requirements are included in the specifications of various organizations such as the U.S. Nuclear Regulatory Commission (NRC), the Department of Defense, transportation authorities, and other large organizations. These specifications include requirements governing the allowable ignition, flame spread, and smoke production of materials, along with the design, installation, and use of electrical devices and systems.

In addition to direct regulation of the flammability and smoke generation properties of materials and equipment, there is also interest in indirect effects on the operational capabilities of personnel and equipment that may be exposed to a fire environment resulting from a fire. In NPP applications, operators must be able to perform appropriate safe shutdown operations and equipment may be critical to plant monitoring or to safe shutdown procedures so that both the direct effects of fire and indirect effects of fire generated smoke are of interest to ensure safe plant shutdown.

Considerable research has been conducted on the direct effects of fire in NPP applications [11]. Recent studies have reviewed potential sublethal effects of fire effluent [13]. A review of both direct and indirect effects of fire on electrical equipment [7] was also a resource for this report. This chapter provides a review of studies of indirect fire effects on electrical equipment including studies specific to NPPs plus a significant base of research related to telecommunications equipment.

2.1 Nuclear Power Plant Equipment

2.1.1 Susceptible Components in NPP Applications

The earliest treatment of indirect fire damage to nuclear power facilities is a report prepared for the NRC by the NUS Corporation in 1985 [14]. The report evaluates the damaging aspects of fire environments, the susceptibility of various components to damage and the importance of those components to plant safety. Of particular interest was the impact of

- conditions associated with suppression activities (high humidity or liquid water effects),
- elevated temperatures below ignition temperatures of typical materials, and
- corrosion due to products from cable fire or gaseous suppression agents.

Components were selected for evaluation based on the Fire Hazards Analysis (FHA) reports from four NPPs. Those components necessary to achieve and maintain safe-shutdown were evaluated based on Environmental Qualification (EQ) test reports and hydrogen burn tests. The EQ reports were used to judge resistance to elevated temperatures and high humidity or liquid water effects. The hydrogen burn tests and numerical simulations subjected a variety of electrical and electromechanical components to elevated temperatures, pressures and humidity caused by hydrogen fires in confined spaces. The hydrogen burn results indicated functionality of specific components and likely causes of failure.

Based on the sources above and extensive use of engineering judgment, components were assigned ratings in a number of distinct categories describing damageability and significance to plant operability. These ratings were weighted according to relative importance and combined into a single value to describe the overall hazard produced by exposing a given component to adverse environmental conditions. The primary result of these experiments is a relative ranking of components by both importance and susceptibility to fire damage. This information, shown in Table 1, is expressed for each component as a single number ranging from 0 to 1, with higher values indicating greater importance and risk of damage. Because the tests used clean burning hydrogen, the applicability of these results is limited with respect to smoke damage, but does provide guidance for identifying components that are both important to plant operation and sensitive to environmental conditions. It is also worth noting that this assessment applies almost exclusively to analog technology.

Table 1. **Relative ranking of NPP electrical components considering both importance and susceptibility to fire damage. Bolded items are ones that were specifically mentioned as being susceptible to moisture, particulates, or corrosion. Data from reference [14].**

Component	Rank	Component	Rank
Recorders	**0.79**	Cables, non 1E	0.33
Logic Equipment	**0.77**	Thermocouples and RTDs	0.33
Controllers	**0.71**	**Pressure Switches**	**0.33**
Power Supplies	**0.67**	Control Transformers	0.29
Meters	0.61	**Motors (open)**	**0.28**
Solid State Relays	**0.60**	Power Transformers	0.26
Electromechanical Relays/Contactors	**0.59**	Position/Limit Switches	0.26
Hand Switches/Pushbuttons	**0.50**	Valve Positioners/Operators	0.25
Transmitters	0.50	Gauges	0.20
MCC	**0.49**	Cables,1E	0.18
Switchgear	**0.49**	**Terminal Blocks**	**0.18**
Battery Chargers/Inverters	0.49	Motors (enclosed)	0.17
Batteries	**0.44**	Fans	0.05
Temperature Switches	0.41	Heaters	0.02
Distribution Panels	0.38	Valves	0.02
Indicating Lights	0.37	Pumps	0.00
Solenoid valves	0.34		

2.1.2 Equipment Exposure to Full-Scale Fire Environments

In response to reports of significant damage caused by so called "secondary environments" created by fires in NPP and other applications, Sandia National Laboratories (SNL) performed a number of cabinet burn tests aimed to better understand the safety issues associated with fires in NPPs [15]. These environments include elevated temperatures and humidity and the presence of particulates and corrosive gases. The primary goal was to determine functionality of components when exposed to these environments. Data were also collected to characterize the environments to which those components were exposed. In addition to the burn tests, a small number of thermal failure and long term corrosion tests were included.

The components used in the tests were selected to represent the most easily damaged NPP electrical components, as identified in Ref. [14]. Some of these were powered during the test and subject to active monitoring, while others were unpowered and evaluated for functionality after the test was complete. In order to add conservatism and represent a wider variety of designs, some components were placed in non-standard orientations or modified (such as by removal of protective cases) to increase susceptibility to expected causes of failure.

Five burn tests were performed, all using unqualified[1] polyvinyl chloride (PVC)-insulated cable as the fuel package. Room size and cabinet configuration were varied as was the arrangement of components. The fires lasted between 15 minutes and 40 minutes, and exhibited peak heat release rates between 185 kW and 995 kW. The fires were allowed to burn completely; no suppression was used.

Component failures and degradation due to the room-scale burn test were as follows:
- Switches exhibited slight sensitivity to fire exposure. For some, a small number of voltage stresses (at most 15 Vac) were required to resume conduction while others had only slight increases in contact resistance. Fire size and exposure were seen as the most significant factors contributing to degradation. Overall, fire exposure did not impede normal operation of the switches tested.
- Relays (powered and unpowered) showed minor signs of corrosion after testing but did not suffer any loss of functionality.
- Meters did not suffer any loss of functionality. It is noted that these are generally well sealed, making them less susceptible to infiltration by products of combustion.
- Pen-based chart recorders suffered mechanical failure due to particulate deposition. There was no indication of electrical failure.
- Electronic counters did not fail during the burn test, despite significant particulate deposition.
- Some power supply and amplifiers responded adversely to increased temperatures, but were not affected by the products of combustion and functioned properly after the test.

In a secondary test procedure, the smoke-contaminated electronic counters were subjected to high humidity while powered. The less contaminated of the two exhibited no loss of

---

[1] The IEEE-383 [16] standard provides requirements for qualifying electrical cables and field splices for electrical service systems used in nuclear power generating stations. The use of non-qualified cable implies that the wiring may not meet the standard requirements and should be considered when assessing the test results.

functionality, while the other failed after an unknown amount of time in a 95 % humidity environment. It was determined that the presence of particulates and moisture had resulted in leakage currents sufficient to blow a fuse in the device. Removal of contaminants from two locations restored functionality.

The most prevalent acid gas released during NPP fires is likely to be Hydrogen Chloride (HCl) since it can be released in significant amounts from burning PVC-containing cables. Active measurements of chloride ions in exhaust ventilation were significantly below expected values, while concentrations in soot deposits were quite high. The theory presented to explain this phenomenon is that the majority of chloride ions were absorbed by smoke particulates in the time it takes for them to reach the exhaust vents. The exhaust ventilation system included a particulate filter upstream of the chloride ion measurement system, thus any ions absorbed by soot would not be measured. It is thought that this phenomenon had not been encountered in previous work because of the relatively short time scales associated with small scale tests, which would not allow a significant amount of ion absorption.

2.1.3   Smoke Exposure Testing

Beginning in 1996, SNL performed a large number of smoke exposure experiments for the NRC. In order to create the greatest level on consistency and repeatability, all of these tests used a single smoke generation and exposure method loosely based on the withdrawn draft ASTM International (ASTM) E05.21.70 corrosivity measurement standard[2]. In this method, fuel samples are exposed to unpiloted ignition by radiant heat (usually at 50 kW/m^2) and the products of combustion are trapped in an enclosed volume containing the components to be tested. When possible, the original ASTM E05.21.70 chamber, consisting of a single combustion furnace supplying a 0.2 m^3 (7.1 ft^3) enclosure, was used. For components larger than the original exposure chamber, a larger, 1 m^3 (35.3 ft^3) enclosure supplied by four combustion furnaces was used. Unlike the draft standard, which specified fuel packages in terms of surface area and volume, the fuel loads for these tests were chosen by weight, in order to produce controlled masses of smoke per unit volume in the chamber. Humidity and temperature were kept at 75 % relative humidity (RH) and 23.9 °C (75 °F) before and after exposure. The exposure period lasted for a total of one hour, with the combustion chambers being active for the first 15 minutes. The chamber includes an optical transmission measurement system, which uses a constant flow of nitrogen to prevent soot deposition onto its optical surfaces. In some tests, this system was deactivated out of concern that the dry nitrogen would artificially lower the humidity of the chamber.

The fuel loads used in these tests were mixtures of a variety of cables that had been identified as being in use in NPPs. To the extent possible, the weight fraction of each cable variety was selected to match frequency of its usage, to create smoke that would generally represent the state of the industry.

The remainder of this section describes testing which used this exposure method.

---

2 See Section 3.2.2.1.2 for additional details on the test method.

2.1.4 Early Digital Equipment Testing for NPP Applications

As digital electronic devices continue to supplant their analog predecessors in NPPs, it becomes necessary to evaluate the ability of the new equipment to perform required monitoring and safety operations while exposed to adverse conditions. A broad study on the impact of environmental conditions on digital systems used in NPPs included an evaluation of smoke effects. Researchers at Oak Ridge National Laboratory and SNL subjected an Experimental Digital Safety Channel (EDSC) to various smoke loads to monitor its functionality [17]. The EDSC is a network of computers and communication devices for monitoring and controlling a nuclear reactor. Exposure tests were performed on individual components of the EDSC while they were functioning and communicating with the EDSC. The behavior of the entire system was monitored for possible failure modes during and after exposure. Due to the size of the EDSC components, the larger 1 m^3 chamber was used.

Three components of the EDSC were exposed to smoke (see Figure 1): the Process Multiplexing Unit (PRS/MUX), the Digital Trip Computer (DTC) and the fiber-optic serial datalinks to and from the DTC, termed a Fiber-Optic Module (FOM). The PRS/MUX served to convert a number of distinct analog electrical signals into a single optical digital signal to be transmitted on the Fiber Distributed Data Interchange (FDDI) ring. The DTC interprets optical digital system measurement information from the PRS/MUX and transmits optical digital trip (reactor shutdown) signals to the Host Processor (HOSTP).

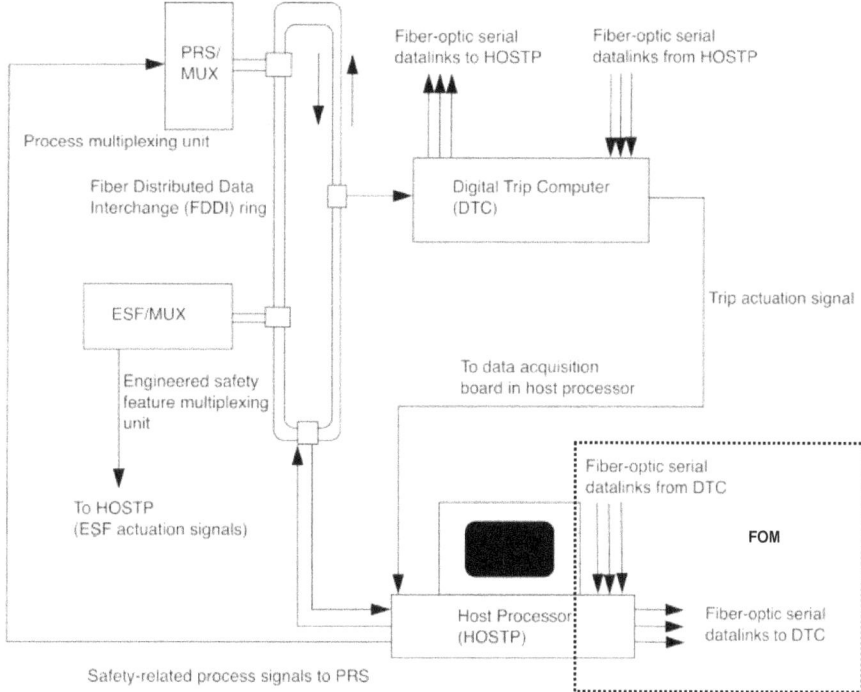

Figure 1.   Functional block diagram of an experimental digital safety channel for NPP applications used to evaluate smoke effects on digital equipment [17].

Based on previous work, nominal smoke loads of 3 g/m^3, 20 g/m^3, and 160 g/m^3 were used to replicate predicted fire conditions. In order to approximate environmental conditions created by suppression activities, some tests introduced steam or CO_2. The introduction of 34 g of steam immediately after the burning of the fuel raised the relative humidity in the chamber to 85 %.

The limited number of tests and the re-use of equipment make it difficult to draw strong conclusions from the results of these tests. No single combination of conditions and equipment was tested more than once, nor were all possible conditions tested. It was noted that pieces of equipment that had been exposed to smoke were no longer able to function without error in a smoke-free environment, despite having been cleaned after each test. As a result it is difficult to distinguish between new and existing damage in the second, third and fourth tests of items.

In most smoke tests (and some baseline tests after initial smoke exposure), a minor communication error requiring that the DTC retransmit data to the HOSTP was recorded. The exact cause of this error was not conclusively determined, although it was suspected that infiltration of particulates into fiber-optic connections or circuit bridging by particulates was to blame.

In two tests, the PRS/MUX was exposed to both smoke and high RH. In both cases, the previously mentioned DTC retransmission error occurred. In the case of low smoke density (3 g/m^2) no other errors manifested. However, during higher smoke density (20 g/m^2), the voltage signals transmitted by the PRS/MUX deviated from those given to it as inputs. No cause for this behavior is stated explicitly, but circuit bridging is implied.

In one DTC test and in the FOM test, timeout errors between the HOSTP and DTC occurred on three datalink channels. In both cases, it appeared that circuit bridging on edge connectors caused the failure. In the FOM test, orientation (vertical or horizontal) had no observable impact on failure. The FOMs were exposed to three fuel packages (2.43 g, 15.45 g and 46.42 g) in succession without venting the chamber or cleaning components. Both failures occurred one hour after the 2.43 g burn, and during the 15.45 g burn.

Tests of CO_2 with or without smoke had no effect on the functionality of the circuits, but did drastically reduce the temperature of the environment.

Overall, SNL concluded that important failure mechanisms included both long-term corrosion and short-term current leakage or circuit bridging, particularly on the typically uncoated edge connections and interfaces. Digital systems not directly exposed to fire were seen to be able to maintain functionality for about one hour following exposure.

2.1.5 Smoke Exposure and Circuit Bridging in Digital Equipment

SNL conducted tests specifically to assess the likelihood and impact of circuit bridging in digital circuits exposed to smoke. Smoke exposure tests were performed on a group of components selected to be typical of modern microprocessor electronics at the time [6].

During the tests, the smoke exposure environments were altered to represent different fire scenarios. The size of the fuel package was varied to either 3 g or 100 g, and in some tests the standard fuel mixture was modified to include PVC insulated cables. Other prescribed deviations from the standard test procedure were applied in various tests. These modifications were reduction of the furnace heat flux to 25 kW/m^2, reduction of the humidity to 20 % RH, application of compressed CO_2 as a post burn suppression agent and inclusion of galvanized sheet metal among the test components. The use of galvanized metal was prompted by reports on corrosion-related phenomena that can result in conductive liquids capable of damaging electronics [6].

The first variety of test board had a pattern of copper traces forming four interdigitated combs. The theory of the test is that deposition of soot and other products of combustion would provide a conductive path between the combs and reduce the overall resistance of the pattern. The initial resistance between the combs varied between $1x10^8$ Ω and $1x10^{15}$ Ω. During the smoke exposure, potentials of 0 V, 5 V, 50 V, and 160 V were applied to investigate the hypothesis that local electric fields could promote soot deposition. The results show the relative change in resistance after exposure.

The second test board measured resistances between adjacent leads on a variety of printed circuit board (PCB) mounted chip packages. These included four surface-mounted components and three through-hole-mounted components. The components represented a range of common geometries using either plastic or ceramic packaging. The boards were energized to 5 V and actively monitored for resistivity changes during smoke exposure. In every test of the chip packages and the comb patterns, four test boards were used; one was kept outside of the exposure chamber as a reference, one was kept in a computer chassis inside the exposure chamber, one was sprayed with a protective acrylic coating and one was completely unprotected.

In addition to resistance measurements, the test boards, dual in-line package (DIP) optical isolator chips and memory chips were tested for functionality. The optical isolators were evaluated by measuring their ability to reproduce a supplied square wave signal, the amplitude, and the delay of response to inputs. The exact nature of tests performed on the memory chips after exposure is not specified, but rather expressed by pass/fail criteria.

The most fundamental evaluation of circuit bridging in these tests were active resistance measurements of PCBs before, during and after fire exposure.

It was observed that comb patterns held at higher voltages collected more soot than identical patterns with low or no voltage applied. Also, when large particles of soot struck the highest voltage patterns (160 Vdc) visible sparks were observed. There were no quantified data presented on this topic.

9

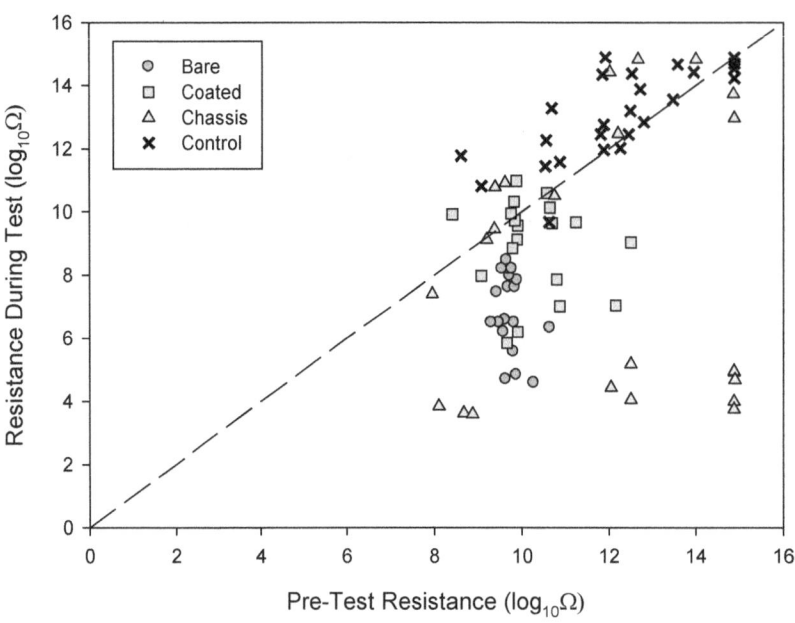

(a) Change in resistance from pretest (X) and during test (Y)

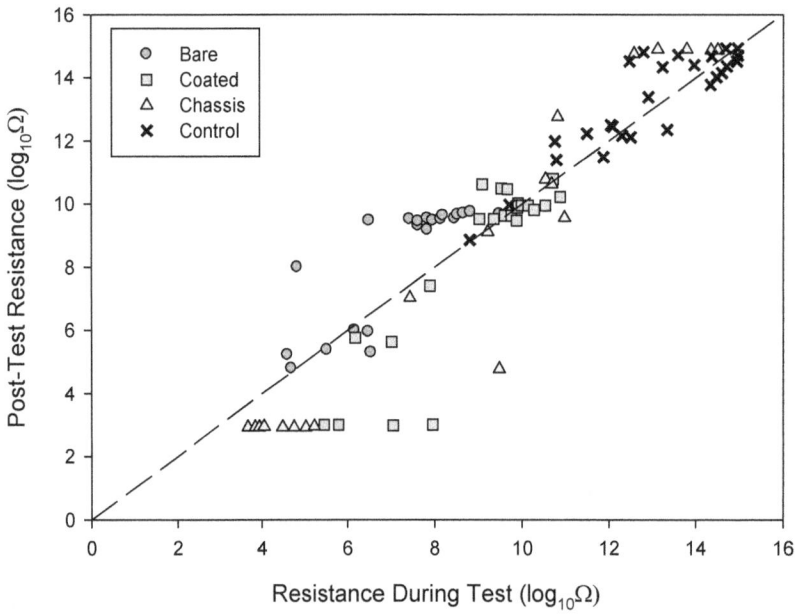

(b) Change in resistance during test (X) and post-test (Y)

Figure 2. Change in resistance for 160 Vdc comb pattern circuit boards exposed to smoke from a fire source. Data from reference [6].

Figure 2(a) shows the resistance before and during exposure for the 160 Vdc comb patterns. The line indicates no change in resistance so that points below it correspond to a decrease in resistance while points above correspond to an increase. Boards exposed to smoke (bare, coated and chassis-mounted) generally showed significant drops in resistance. Some sense of the consistency of the experiments can be seen in the results for the control boards, which ideally would have exhibited constant resistance but, in fact, increased in resistance. These results are representative of data from testing as a whole. Similar resistance changes occurred for lower voltage combs and for components mounted to the boards.

Figure 2(b) presents comparable results for resistance during exposure and averaged over the first 22 hours following venting of the chamber. These data provide insight into the tendency of components to either recover from damage or suffer long term degradation. Components which initially maintained a relatively high resistance tended to recover slightly while those that showed large resistance losses during exposure continued to suffer damage after venting.

In order to determine the significance of test parameters, linear models were constructed to predict the resistance before, during and after exposure. Individual factors as well as combinations of two and three factors were expressed as binary values and assigned coefficients by means of a least-squares regression analysis. Since insignificant factors (those having very small coefficients) were omitted, the frequency with which a factor or group of factors appears in these models is a strong indicator of the impact (either positive or negative) that it has on circuit bridging. The condition of the board (bare, coated or chassis mounted) and humidity level were seen to be the most influential individual factors. It is also clear that bare boards were very susceptible to a combination of high fuel load and high humidity; while chassis mounted boards were heavily influenced by the combination of high fuel load and the presence of galvanized metal in the test chamber. Although the presence of PVC in the fuel did not appear to be significant, the authors note that the basic cable did contain high concentrations of Cl and Br and may have had a similar impact on conductivity and corrosivity of gases.

In 4 of the 27 tests, the isolator stopped transmitting a signal entirely. In other tests, there were significant differences between input and output signals, which could adversely affect the function of systems that depend on those signals. The memory chips failed in 6 of the 27 tests. In all but one case, failures of these components occurred only when the previously discussed linear models predicted low resistance across the leads of DIPs.

These tests clearly demonstrate the potential of smoke deposition to cause failure of electronic systems by circuit bridging. The humidity and amount of fuel consumed both strongly contributed to the degree of bridging. Also, flaming rather than smoldering combustion was more likely to result in lowered resistance. Conformal coatings and mechanical enclosures were seen to provide a measure of protection from smoke.

2.1.6   Effects on Smoke on Functional Circuits

SNL also tested complete and functional circuit boards for their susceptibility to damage from smoke [18]. These boards contained five functional circuits that could be energized and monitored during smoke exposure. Each circuit was present twice on each board, built with either surface-mount technology (SMT) or pin through hole (PTH) components. Half of the boards tested were protected with a polyurethane conformal coating. The circuits included the following:

11

- High-Voltage Low-Current (HVLC): This high-resistance series of capacitor-resistor pairs was subject to a 300 Vdc bias. The nominal current was 6 µA. Within the network itself, the potential difference between any two adjacent copper traces was 60 Vdc.
- High-Current Low-Voltage (HCLV): This circuit contained a number of resistors and capacitors in parallel. A constant-current power supply adjusted the applied voltage to maintain 1 A through the nominal 1.43 Ω.
- High-Speed Digital (HSD): This circuit simply connected a series of NAND logic gates. While powered, a 5 V, 20 ns input pulse was applied to the first gate, and the output of the final gate was monitored. Increases in the rise, fall and delay (relative to input) times of the output square pulse all indicate degradation of a circuit.
- High-Frequency Low-Pass Filter (HFLPF): a system of inductors and capacitors configured to attenuate high frequency input signals. Before and after exposure, the filters were tested for signal attenuation from 50 MHz to 1 GHz. During exposure, they were tested for: attenuation at 50 MHz, 250 MHz and the frequencies at which attenuation reached -3 and -40 dB.
- High-Frequency Transmission Line (HFTL): this circuit was two adjacent PCB copper traces, designed to exhibit 50 Ω impedance. It is expected that a small amount of a signal on one line will be transmitted to the other. This behavior, referred to as coupling, is frequency dependent and is expected to increase with the conductivity between the lines. During exposure, coupling between the lines was measured at 50 MHz, 500 MHz and 1 GHz. Pre- and post-exposure tests added measurements of coupling with the second line connected in a reversed configuration.

In addition to the functional circuits, additional components on the board were tested for leakage currents before, during and after exposure. Further leakage current measurements were performed using an interdigitated comb test board. Each board had four distinct comb patterns and during exposure, two were biased with 5 Vdc and the other two with 30 Vdc.

The tests used the standard exposure method, with fuel loads of 3 g/m^3, 25 g/m^3 and 50 g/m^3 heated with 50 kW/m^2 and 25 kW/m^2. Soot deposition was measured with two quartz crystal microbalances (QCM), one oriented vertically and the other oriented horizontally. The quantities deposited and fuel burned are listed below in Table 2. Note that the 25 kW/m^2 heat flux exposures caused smoldering of the fuel while the 50 kW/m^2 flux caused flaming.

Table 2. Soot deposition during smoke exposure tests of functional circuit boards for NPP applications. Data from reference [18].

Fuel level (g/m^3)	Fuel available (g)	Flux level (kW/m^2)	Fuel burned (g)	Vertical deposition (µg/cm^2)	Horizontal deposition (µg/cm^2)	Deposition ratio (V/H)
3	6	25	0.27	0.4	2.8	0.14
3	6	50	0.49	0.9	3.4	0.27
25	5	25	3.12	2.1	45.3	0.05
25	5	50	3.95	4.2	17.6	0.24
50	10	25	6.49	3.2	94.8	0.03
50	10	50	7.34	7.3	58.5	0.13

Chemical analysis of soot was performed on deposits collected by 14.2 cm^2 disks of ashless filter paper on the floor of the test chamber (Table 3). With the larger fuel packages, smoldering resulted in larger quantities of deposited chloride and bromide.

Table 3. Concentration of chloride, bromide, and fluoride ions during smoke exposure tests of functional circuit boards for NPP applications. Data from reference [18].

Fuel level (g/m^3)	Flux level (kW/m^2)	Chloride (µg/Filter)	Bromide (µg/Filter)	Fluoride (µg/Filter)
3	25	95.5	2.1	0.0
3	50	24	5.5	0.25
25	25	92.6	96.6	0.0
25	50	73.0	66.9	1.45
50	25	147	181	1.25
50	50	95.4	165	0.0
Unexposed filter	0	3.9	0.0	0.7

No direct measurements of corrosion were performed, but visual inspection of components showed that components made of reactive metals (such as nickel) had corroded significantly after the test.

The HVLC circuit showed susceptibility to smoke exposure. Resistance decreased noticeably with smoke density and recovered somewhat after venting of the chamber. Conformal coatings were very effective at preventing or limiting resistance losses for all fuel loads. In this particular circuit, the PCB traces were the most vulnerable to damage because they carried the highest potential differences and were more closely spaced than the leads of attached components. Because this vulnerability was independent of the types of components attached, there was no observed difference between PTH and SMT.

The HCLV circuit was largely insensitive to smoke. Because the circuit had a very low initial resistance, the addition of weakly conductive alternate paths would have little impact. The exception was the case of the bare (having no conformal coating) SMT board. In all of the medium and high fuel load tests, the resistance of these circuits increased measurably (0.9 % -

1.7 %) and did not recover after venting. This change is attributed to corrosion of contacts or solder joints. None of the changes in these tests represented a significant effect on system performance.

The HSD circuit was tested for its response to a 20 ns, +5 Vdc pulse by measuring the output rise time, fall time and delay. Changes in signal degradation due to differences in cable length make direct comparison between pre-test results and *in situ* measurements invalid. However, none of the circuits showed significant change in response to changing fuel loads or the presence of conformal coatings. Previous work indicates that the highest fuel load used would have caused failure due to circuit bridging in complementary metal-oxide semiconductor (CMOS) chips, but that advanced Schottky transistor to transistor logic (FAST TTL), such as those in the HSD, are less sensitive.

The HFLPF did not show any response to smoke. However, active measurement during exposure was performed with an input frequency (250 MHz) that would not be impacted by changes in system capacitance due to debris (the expected failure mode). Post-test measurements over a wider range of frequencies showed no change from the pre-test condition. It was expected that the most severe changes in capacitance would occur during exposure, when the testing method would be unable to measure those changes.

The HFTL circuit was tested to determine coupling between two adjacent lines. Coated boards showed no change in coupling during or after exposure. The bare board experienced a large (from -52 dB to -17 dB) increase in coupling of 50 MHz signals during exposure. Coupling on the bare board returned to pre-test values quickly after venting of the test chamber. At high frequencies (500 MHz and 1 GHz) coupling on the bare board actually decreased during smoke exposure and returned to pre-test values after venting.

Leakage current data are presented only for the PTH boards and the interdigitated comb; leakage data for other components in the methodology section were not reported. Resistance on the bare boards dropped as fuel load increased and recovered somewhat after venting. Changes in the interdigitated comb were less consistent, but surface resistance generally decreased with increasing fuel loads and partially recovered after venting. Results in Table 4 are presented in the change of $Log_{10}(R)$ between pre-test values and average during smoke exposure.

Table 4. **Change in resistance from smoke exposure of an interdigitated comb circuit board exposed to fire smoke. Data from reference [18].**

Fuel(g) / Flux(kW/m^2)	0/50	3/25	3/50	25/25	25/50	50/25	50/50
$\Delta Log_{10}(R)$ boards	0	0	-1.5	-6.1	-6.3	-6.5	-6.7
$\Delta Log_{10}(R)$ comb (5V)	0.6	0.7	0.3	-0.6	-1.5	-2.6	-1.9
$\Delta Log_{10}(R)$ comb (30V)	1.2	0.1	-1.9	-0.2	-3.2	-1.1	-2.8

2.1.7 Other Smoke Effects Testing

In a final report in the series of non-thermal damage experiments, SNL reports on previously unpublished results including tests of conformal coatings, digital throughput, memory chips, hard disk drives, and electrical properties of fire smoke [19].

2.1.7.1 Conformal Coatings

In order to gauge the ability of conformal coatings to mitigate the effects of smoke exposure, a battery of high smoke load (200 g/m^2) exposure tests were performed on functional boards with a variety of conformal polymer coatings (listed in Table 5, below) [19]. The circuits tested were the previously discussed HVLC, HCLV, HSD, HFTL and HFLPF constructed from PTH and SMT components. There were variations in response with different coating chemistries and electrical components, but in general, coated boards were less susceptible to damage than bare boards.

Table 5. Conformal coatings applied to functional circuit boards to mitigate the effects of exposure to fire smoke. Data from reference [19].

Coating Type	Brand[3]	Product	Thickness (mils)	Application Method
Acrylic	Humiseal	1B-31	2.5	Dipped
Epoxy	Envibar	UV1244	2.5	Dipped
Parylene	Union Carbide	Type C	0.75	Vacuum Deposited
Polyurethane	Conap	CE-1155	2.5	Dipped
Silicone	Dow	3-1765	5	Dipped

When constructed from PTH components, the HVLC circuit maintained its resistance best with parylene or polyurethane coatings and varied the most (excepting the bare board) with the silicone coating. The SMT components maintained resistance well with parylene, polyurethane and acrylic coatings. Although the expected and most commonly observed degradation mode in the HVLC circuit was a loss of resistance, some PTH components showed increases in resistance during the burn phase of testing. It should also be noted that the epoxy coated SMT board did not recover appreciably after smoke was vented from the chamber, a behavior which was not observed in any other configurations (including bare boards).

In the HCLV tests, only the bare SMT board showed a noticeable change. After the burn began, the resistance of the circuit rose from 1.46 Ω to 1.49 Ω. Resistance continued to increase slowly until the chamber was vented, at which point it remained constant. The bare SMT board showed no signs of recovery. With such a small change in resistance, it is not clear that this is a significant result.

---

[3] Certain commercial entities, equipment, or materials may be identified in this document in order to describe an experimental procedure or concept adequately. Such identification is not intended to imply recommendation or endorsement by the National Institute of Standards and Technology, nor is it intended to imply that the entities, materials, or equipment are necessarily the best available for the purpose.

The HSD was protected very well by all conformal coatings. Of the 18 bare boards (9 SMT and 9 PTH), 7 PTH boards and one SMT board experienced momentary transmission failures at some time after testing. Also, post test measurements showed that fall time response on the bare PTH increased from 1.6 ns to 2.2 ns, while SMT and all coated boards remained unchanged.

As observed in previous tests, the bare transmission line circuits showed increased forward coupling at 50 MHz and decreased forward coupling at 500 MHz and 1 GHz while smoke was present. Coated boards showed no coupling response to smoke.

The bare HFLPF circuits were mildly affected by smoke exposure, while coated boards showed almost no response.

2.1.7.2   Digital Throughput

Smoke exposure tests were performed with a variety of fuels to determine the impact of smoke exposure on digital cable connectors [19]. A circuit board with linked pairs of D-subminiature 9-pin (DB-9), D-subminiature 25-pin (DB-25) and network modular (RJ-45) connectors was placed in the exposure chamber. Digital signals from a personal computer outside of the chamber were routed through this board in order to determine the impact of smoke upon these connectors. PVC, Douglas fir and jet fuel were used as fuels for these tests, rather than the standard cable mixture.

DB-9, DB-25 and RJ-45 connectors were tested simultaneously in 19 tests and showed no signs of failure. Subsequent tests performed by intentionally bridging signal lines with a variable resistance indicate that conductance between pins on a connector must closely approach the conductance of the cables themselves before there is a significant risk of failure.

2.1.7.3   Memory Chips

Smoke exposure tests were performed on memory chips to determine functionality and attempt to isolate physical parameters that can predict chip failure [19]. The standard exposure method and fuel mixture were used, with additional electrical testing continuing for one hour after venting the chamber. Smoke loads of 30 g/m^3, 35 g/m^3, 50-60 g/m^3, and 150 g/m^3 were used. The five types tested were 3.3 V volatile Static Random-Access Memory (SRAM), 5 V volatile SRAM, DIP Erasable Programmable Read-Only Memory (EPROM), plastic-leadless chip carrier (PLCC) EPROM and 5V non-volatile SRAM.

Functional testing of the chip consisted of repeatedly writing data to the chips and reading it back. The response time of the chips and the error rate of read/write operations were recorded. The parametric measurements monitored leakage current behavior. The current drawn when the chip was powered but not operating, current when a standard voltage is applied to a given pin and the voltage required to maintain a particular current were measured repeatedly at 30 s intervals.

The exposure tests for electronic memory determined both functionality and the electrical parameters (voltage, current etc.) of the chips during and after smoke exposure. Functionality was determined by the ability of the chip to be accurately written to and read from. Table **6** indicates success and failure rates. The non-volatile SRAM suffered no failures even at higher (150 g/m^3) fuel loads.

Table 6. Results of memory chips exposed to fire smoke. Data from reference [19].

Fuel Load	30 g/m^3 Pass/Tested	50- 60 g/m^3 Pass/Tested
3.3 V SRAM	2/2	0/4
5V SRAM	2/2	2/4
DIP EPROM	1/1	5/5
PLCC EPROM	1/1	5/5

Comparison of failures to the parametric data showed no consistent correlation between leakage current and failure. When the 3.3 V SRAM chips failed, their operating current roughly doubled, while the operating current of the 5 V SRAM dropped to half of normal during failure. Although some of the recorded failures occurred shortly after the smoke chamber vented, all chips functioned properly when tested several weeks later, indicating recovery after exposure.

2.1.7.4 Hard Disk Drives

Hard disk drives (HDD) isolated from other computer components were tested for read/write functionality during and after exposure and were scanned for corrupted data sites after exposure [19]. Five drives were tested six times. In the first test the drives were oriented as they would be in normal use. In all subsequent tests, the drives were inverted to increase soot deposition onto exposed circuit boards on the underside of the drives. The standard smoke exposure method was used with fuel loads of 15 g and 30 g. In two of the tests, the nitrogen purge protecting the optics was deactivated, to prevent the dry nitrogen from lowering the humidity of the chamber. All drives passed the *in situ* functionality tests. One of the 5 disks showed signs of corrupted data, but the cause was not identified.

2.1.7.5 Electrical Measurements

Two important electrical properties of smoke are conductivity in the gas phase and conductivity on surfaces. Surface conductivity measurements were performed with the previously described interdigitated comb and a specialized low-mass comb board. Gas conductivity was measured with pairs of freestanding, parallel perforated stainless steel plates. Masses of fuel between 1.5 g and 19 g were used.

The comb patterns were biased with 5 V, 50 V and 500 V during exposure. Current flow was measured continually during and after exposure. The boards were weighted and visually inspected after exposure, to identify the mass and physical distribution of the deposited smoke.

In one set of tests [19], the vertical plates were placed with a fixed spacing of 2.5 mm. In these tests, the plates were biased with 5 Vdc, 50 Vdc, 500 Vdc and 1 Vac. The current flow between the dc-biased plates was monitored during exposure. The ac-biased plates were analyzed to measure admittance from 500 kHz to 30 MHz. Video of these experiments was recorded to monitor the behavior of soot particles in electrical fields.

In a different set of experiments [19], the plates spaced at distances from 3 mm to 25 mm subjected to a rising ac bias, up to 4.2 kV. The voltage was allowed to rise until arcing occurred. Current and voltage behavior during arcing were measured.

It was found that higher voltages (50 V and 500 V) attracted smoke particles to the plates and significantly increased conductance between the plates. The smoke particles were observed to accumulate between the two plates and create conductive bridges. As a result, conductance increased very quickly with smoke density but decreased much less quickly and leveled out at approximately 20 % of its peak value. Conductance only returned to near zero values when forced ventilation of the chamber disrupted the accumulated soot bridges.

The interdigitated comb boards biased at 5 V, 50 V and 500 V were tested for conductance over time and total mass of smoke deposited. Inspection of the boards after exposure showed that soot preferentially deposited on the traces of the boards at higher voltages, resulting in a less even distribution of soot on the 50 V and 500 V boards. Although distribution pattern varied with voltage, the total mass deposited depended only upon the total fuel load. Conductance increased sharply with the smoke level and decreased as the smoke cleared. The conductance of the 5 V board recovered far less effectively than the 50 V and 500 V boards. It was suggested that this was due to the more even distribution of soot.

2.2 Telecommunications Equipment

Outside of the nuclear power industry, the overall cost of a fire may be more important than a temporary loss of functionality. While the short term failure of equipment has a cost in lost business, the damage caused by non-thermal fire environments can easily total millions of dollars [3]. Thus, the cost of a fire event can be significantly decreased if components exposed to smoke can be recovered and returned to service, rather than being replaced. As such, much of the research in this industry has focused on understanding and mitigating the forms of non-thermal damage that can decrease the lifetime of equipment.

2.2.1 Corrosion

The most commonly identified and studied mechanism of non-thermal damage is corrosion caused by the chemical products of combustion [20 - 23]. Starting in the late eighties, prompted by reports of extensive corrosion damage and accumulated corrosive products found in a number of separate telecommunications facilities during the 1970's and 1980's, researchers began to develop the theoretical and empirical basis for characterizing the release, transport and impact of corrosive compounds from fires.

A significant part of this work was the development of a number of bench-scale laboratory experimental testing methods for determining the potential of the combustion products of a particular fuel to corrode metal (detailed in section 3.2). However, these standards are of a very simplistic nature and do not necessarily provide immediate insight into the complex phenomena associated with real-world fire scenarios. In order to develop prevention and recovery procedures, a number of intermediate and full scale test fires of a more realistic nature were required.

The first steps that Tewarson and Chu [20] took in evaluating the non-thermal damage potential of cables used the Factory Mutual Research Corporation's (FMRC) 50 kW-scale flammability apparatus to evaluate ignition and flame spread characteristics as well as yields of products. This device subjects a vertically oriented sample to radiant heating and a controlled flow of gas at the component sample. The composition of the effluent is analyzed to determine chemical composition, which is used to determine corrosive properties and heat release rate. The separation of chemical product distributions and fire properties allows for a more detailed understanding of the factors that contribute to corrosive damage. Figure 3 shows the HCl yields (g/g) relative to HCl concentrations (g/m^3) in the effluent stream for a selection of PVC cable insulation materials. The weak correlation between the two data sets is a result of the burning rate of the fuels and the dynamic nature of this test method. The forced flow ensures a relatively constant mass flow of air, while the mass release rate of HCl depends on the burning rate of the material. Thus, materials rich in chlorine will produce relatively low concentrations in the gas phase if they burn slowly. Further, the flame propagation rates are seen to play a key role in determining the amount of fuel that will become involved before intervention occurs in a fire

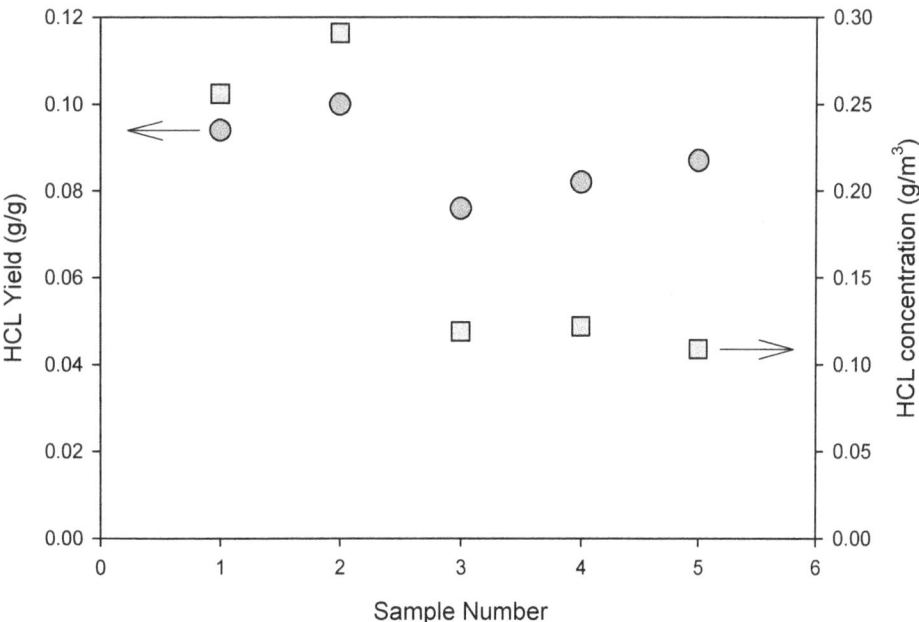

Figure 3. HCl yields relative to HCl concentration for several samples of burning PVC cable insulation materials. Data from reference [20].

19

event. Again, a chlorine rich material may prove to be less of a corrosion hazard than a material with only modest chlorine levels if it is less flammable. This illustrates the importance of considering the interaction of a complete range of physical behavior when determining the meaning of a parameter such as chlorine concentration.

The prevalence of corrosive compounds in telecommunications fires is often a result of halogenated compounds used as solid fire retardants and gaseous suppression agents [24]. When fires involving these compounds do occur, significant quantities of the hydrogen halides and halogen radicals may be released along with the other products of combustion. Therefore, the use of these compounds to reduce the risk of a fire may carry with it the potential to increase the severity of non-thermal damage caused by any fire that does develop [24].

A commonly identified corrosive reaction is that of hydrogen chloride with zinc. [21,22] This has been observed in a number of fires, because of the prevalence of galvanized (zinc coated) steel in HVAC systems as well as support structures in electronic equipment. The solid zinc chloride adheres well enough to the base metal that it poses little direct hazard to electronics. However, the $ZnCl_2$ is extremely hygroscopic, being able to absorb water in relative humidity as low as 10 %. The $ZnCl_2$ is very conductive and can easily flow or drip onto nearby electronics, causing circuit bridging.

2.2.2 Comparison of Fire Performance and Corrosion

Work by Chapin et al. [25] evaluated the ability of a variety of standard test methods to evaluate the impact of smoke from local area network (LAN) cables on electrical equipment. Their concern was that the specialized and controlled nature of these tests may prevent them from realistically predicting corrosion and, more importantly, overall functionality of electronics exposed to smoke. They examined correlation among the standard tests and provided a discussion of the sensitivities and possible sources of error.

Although the discussion and findings address a larger number of tests, the authors performed experiments with only three previously existing test methods. ASTM test D5485, International Standards Organization (ISO) test DIS 11907-3 and International Electrotechnical Commission (IEC) test 60754-2 are described in section 3.2 of this report. These tests burn complete products, individual conductors, or raw materials (individual components of wire insulation) to classify their tendency to cause corrosion. Both the ASTM and ISO tests subject electrically conductive copper corrosion probes to the smoke produced. The change in conductivity of these copper surfaces is used to estimate the amount of metal lost, and results are expressed as thickness consumed (ASTM) or percent consumed (ISO). The IEC test measures pH of gases produced in the burning of a material. Note that the IEC test evaluates individual materials, and as such, cable coatings must be separated into their jacket and insulation materials for isolated testing.

To complement results from the corrosivity standard, other measurements were made to evaluate the fire behavior of the cable samples. The chemical compositions of the cable jacket and insulation materials were determined. Each variety of cable was subjected to the ASTM E1354 cone calorimeter. This uses the same combustion apparatus as the ASTM D5485, but produces

20

heat and smoke release measurements rather than capturing corrosive gasses. This information is used in more traditional analysis of flammability and fire hazards. The smoke release data provide interesting context for the corrosivity measurements by allowing one to evaluate the likelihood and severity of fires involving those materials.

The authors report that, in the course of previous work, they have observed that smoke most often causes the failure of digital electronic systems by decreasing surface resistivity. In an effort to quantify this behavior relative to the corrosivity tests, the authors developed a test apparatus to measure changes of surface resistivity of a circuit board exposed to smoke. The tube furnace used in the IEC 60754-2 test was fitted with an exposure chamber into which the products were driven by forced air flow. The chamber was used to expose printed circuit boards with an interdigitated comb pattern (similar to those used by SNL for the circuit bridging tests described in Section 2.2). After exposure, the targets were tested for current flow with an applied 50 V potential. Leakage currents were tested in a controlled humidity chamber over a range of relative humidity from 30 % to 90 %.

Seven commercially available LAN cables were used in these tests. Three of the varieties were rated for plenum use according to Underwriters Laboratories (UL) test 910, indicating a very low fire hazard. The other cables were rated for other use under different UL and IEC tests. All of the plenum rated cables and some of the non-plenum cables contained large quantities of halogens, which are often cited as a major contributing factor in corrosion caused by smoke [20 - 22].

For the purposes of comparison, results have been normalized by the average value from each test method, and the pH measurements for the IEC 60754-2 have been inverted such that higher values indicate higher acidity. Comparisons of the test results and the correlation between tests are shown in Table 7 and Figure 4 calculated as correlations coefficients from the data in reference [25].

Table 7. Correlation coefficients comparing test results from selected tests of LAN cables. Data from reference [25].

	Total Smoke	ASTM D5485	Jacket pH	Insulator pH	ISO DIS 1197-3	Leakage
Total Smoke	1	0.707	0.274	-0.222	0.193	0.802
ASTM D5485		1	0.551	-0.023	0.728	0.377342
Jacket pH			1	0.734	0.519	-0.095
Insulator pH				1	0.297	-0.369
ISO DIS 1197-3					1	-0.056
Leakage						1

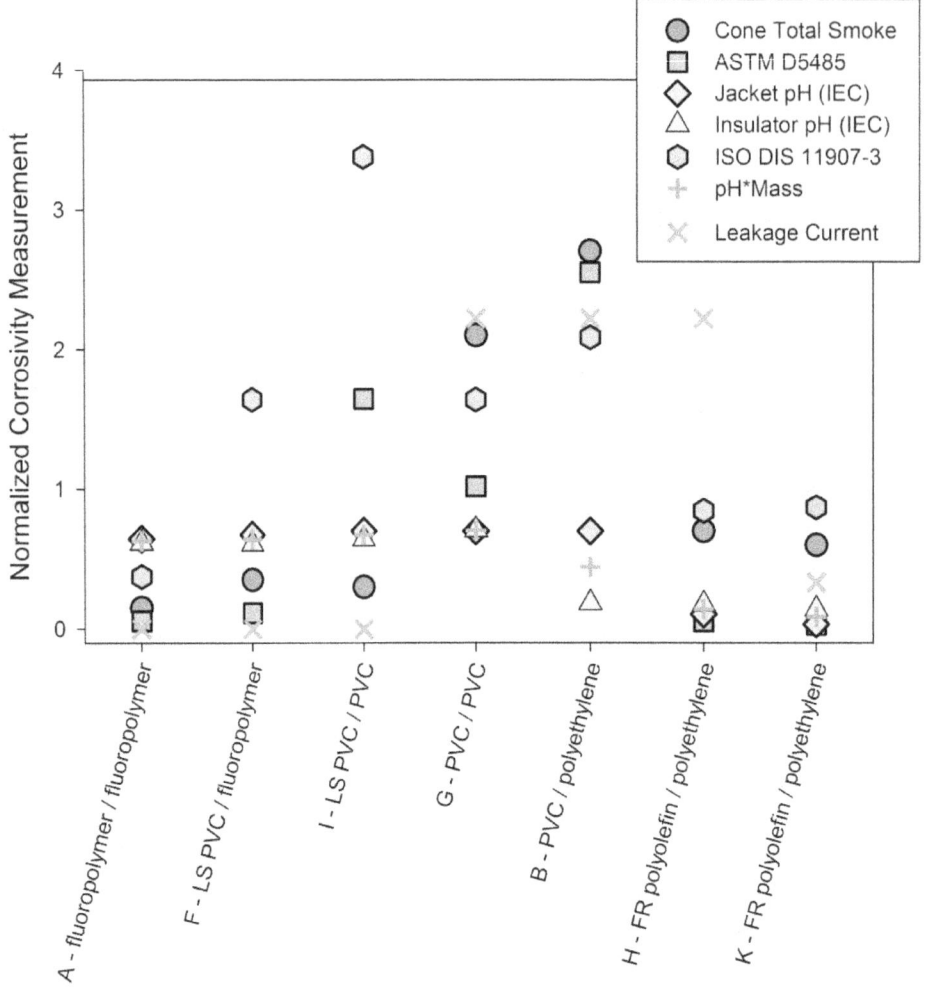

Figure 4. Test results from fire and corrosivity testing of seven commercially-available LAN cables. Data taken from [25].

The plenum-rated cables (cables A,F,I) are easily differentiated from the non-plenum cables by the leakage current test. Leakage currents caused by the plenum cables were 3 to 7 orders of magnitude lower than the highest performing non-plenum cable (K). It may also be possible to make such distinctions based upon smoke load, which was higher for non-plenum cables. The correlation between smoke load and leakage current is consistent with the premise that leakage currents are the result of circuit bridging by particulates in smoke, as has been found by Chapin *et al.* [25] and Tanaka [18].

This low smoke load produced by plenum cables is not surprising, as UL 910, which is based on the method developed by Steiner [26], includes criteria for smoke production as well as heat release rate (HRR) and flame spread. UL 910 was specifically designed to limit smoke production from cables in a plenum configuration [27]. Products that pass can be expected to produce very small amounts of smoke.

The simple calculation of correlation coefficients could suggest that ASTM D5484 is sensitive to smoke load. However, a more careful examination of material composition would indicate that the presence of PVC is the more important factor. The three cable products (cables I,G,B) which caused significant corrosion in this test all used PVC in their jacket and/or insulation, while the other cables did not. It is interesting to note that cable F contained a low smoke PVC-based compound but produced negligible corrosion in ASTM D5484. The amount of smoke actually produced by this cable was slightly higher than the highly corrosive cable I. As both contained similar amounts of chlorine, this suggests that the mechanism used for smoke suppression may also serve to decrease the corrosivity of the combustion products. A larger sample size and a more detailed treatment of the chemistry involved would be required to develop a full understanding of any role that fire retardant compounds may play in corrosion.

Results of the ISO DIS 11907-3 test are more difficult to explain. The three cables which caused corrosion in the ASTM D5484 (cables I,G,B) did so in this test as well. The smoke suppressed PVC cable (F), was comparable to one of the ordinary PVC cables (G). Further, the two non-halogenated cables caused between 25 % to 50 % as much corrosion as the PVC cables in this test as opposed to less than 10 % in the ASTM test. Chapin *et al.* [24] argue that the corrosion target used in the ISO test may produce inaccurate results, due to circuit bridging. The target tracks material loss in the form of increased resistivity in a closely packet serpentine copper trace on a circuit board. They express concern that connections across the board could decrease effective resistance, resulting in underestimation of total mass loss.

The only consistent behavior from the IEC 754-2 pH tests was that halogenated polymers produced low pH, while non-halogenated polymers did not. There was no strong correlation to either of the direct corrosion tests.

None of the corrosivity tests correlated well with one another or with the leakage current test. As such, it is difficult to draw any specific conclusions as to the quality of particular test. The lack of agreement between generally similar tests (such as ASTM and ISO) serves to illustrate the need to understand complex nature of corrosion phenomena in order to properly assess the threat posed by smoke. While Chapin *et al.* [24] claim that circuit bridging is the most important factor in the failure of electronics, others have presented cases in which corrosion dominated equipment losses [20,21].

2.3   Damage Criteria for Electronic Equipment

Early reports on non-thermal damage to telecommunications were based on observations of the methods required to recover equipment from fire events. These reports categorize the level of damage to equipment based on the acid gas concentration on component surfaces. Both Tewarson and Reagor report that deposited mass of chloride can be a predictor of corrosive damage. However, the thresholds specified differ [20,21]. Reagor categorizes four levels of contamination by mass of chloride deposited per unit area: clean at less than 2 $\mu g/cm^2$, easily cleaned and recovered below 30 $\mu g/cm^2$, difficult but possible to recover between 30 $\mu g/cm^2$ and

90 $\mu g/cm^2$, and above 90 $\mu g/cm^2$ it is increasingly likely that the cost of cleaning the component will be greater than simply replacing it. Initially, Tewarson specified a minimum concentration of 0.98 $\mu g/cm^2$ coupled with relative humidity greater than 30 % at 20 °C (68 °F) and gas concentrations of 100 ppm HCl or 1000 ppm of NO_2, HF, SO_2 or acetic acid. In subsequent literature, Tewarson cites Reagor's criteria [28].

Tanaka and Tanaka et al. [18,29] studied failure criteria for digital logic circuits exposed to smoke. In order to evaluate the loss of functionality of the logic circuits, the loss of resistance due to smoke was simulated with a variable shunt circuit to determine the point of failure for a range of circuits. Components with a high tolerance to resistance loss are seen to be more tolerant to smoke exposure. Critical resistance at failure varied for different chips designs, but was highly correlated to the output current of the device. Higher current outputs were seen better able to supply the necessary current to maintain a voltage level despite the loss of resistance due to the shunt.

3 MEASUREMENT OF SMOKE PRODUCTION

During the 1940s and the 1950s, flammability (or "reaction-to-fire") tests were developed on a purely ad hoc basis. Results were typically expressed by arbitrary 0 to 100 scales or by such rating terms as "self-extinguishing." In 1973, the U.S. Federal Trade Commission saw such practices as misleading and sued a number of plastics manufacturers and also the American Society for Testing and Materials (ASTM) [30]. A consent agreement was eventually reached whereby a Bunsen burner test, ASTM D 1692, was dropped, and a caveat was inserted into other ASTM tests, in an attempt to avoid their future misuse. It is noteworthy that the situation in other countries is similar to the U.S. experience. More than thirty years ago, Emmons obtained the results of flammability tests on a number of materials, when tested according to various national, bench-scale flammability standards [31]. He found that the relationship between the test results and real-scale fire behavior according to the different standards was almost completely random. In 1987, Östman and Nussbaum re-examined this issue; the situation appears to have improved only slightly [32]. The reason is that the new knowledge gained in fire physics and engineering over the last decades has generally not yet been reflected in many of the required testing standards.

This is due at least partly to the great number of national test methods that exist for fire testing. For example, one such compilation which includes only the ASTM fire test methods [33] tabulates some 167 test standards covering a broad range of measurements. Based on this large number of tests, it may seem that fire test methods are highly refined and well tuned to specific areas. Though many of the currently published methods were developed up to 40 years ago, for smoke production and for most other fire test results, the focus has shifted to appropriate input data for predictive modeling methods. This chapter provides a review of both types of test methods use to quantify smoke production, those that primarily provide a ranking of materials, and those that provide quantitative input for modeling calculations.

3.1 Smoke Opacity Testing

Typically, bench-scale test methods designed to measure the impact of fire smoke provide a measurement of opacity, an indication of visibility through the smoke. The most significant body of work in this area has been accomplished by Jin [34–36], who found that there is an approximate reciprocal relationship between smoke opacity and visibility distance (the distance at which a person can identify an exit sign, for example), according to

$$kV = 2 \qquad (1)$$

where k is the smoke extinction coefficient (m^{-1}) and V is the visibility distance (m). Thus, measurement of smoke opacity or obscuration is useful in determining the ability to escape or perform needed action in a space during a fire incident. However, opacity is the result of the generation of the smoke and not a direct measure of its generation. Opacity results not only from the generation of smoke, but also from agglomeration and surface deposition during transport. Still, there are a number of widely-used tests that measure smoke opacity.

Smoke opacity tests are broadly categorized as either static tests (those in which the smoke generated is allowed to accumulate within the test enclosure) and dynamic tests (those in which there is a continuous flow of gases resulting from combustion of the sample).

The Smoke Density Chamber (ASTM E 662) [37], is used widely in testing of materials. Outlined in Figure 5, it is a static test that measures smoke generation from small, solid specimens exposed vertically to a radiant flux level of 25 kW/m^2 in a flaming (piloted ignition) or non-flaming mode. The smoke produced by the burning specimen in the chamber is measured by a light source – photometer combination. The attenuation of the light beam by the smoke is a measure of the optical density or "quantity of smoke" that a material will generate under the given conditions of the test. Two measures are typically reported. D_S is an instantaneous measure of the optical density at a particular instant in time and is expressed as

$$D_S = \frac{V}{AL} \log\left(\frac{I_0}{I}\right) \tag{2}$$

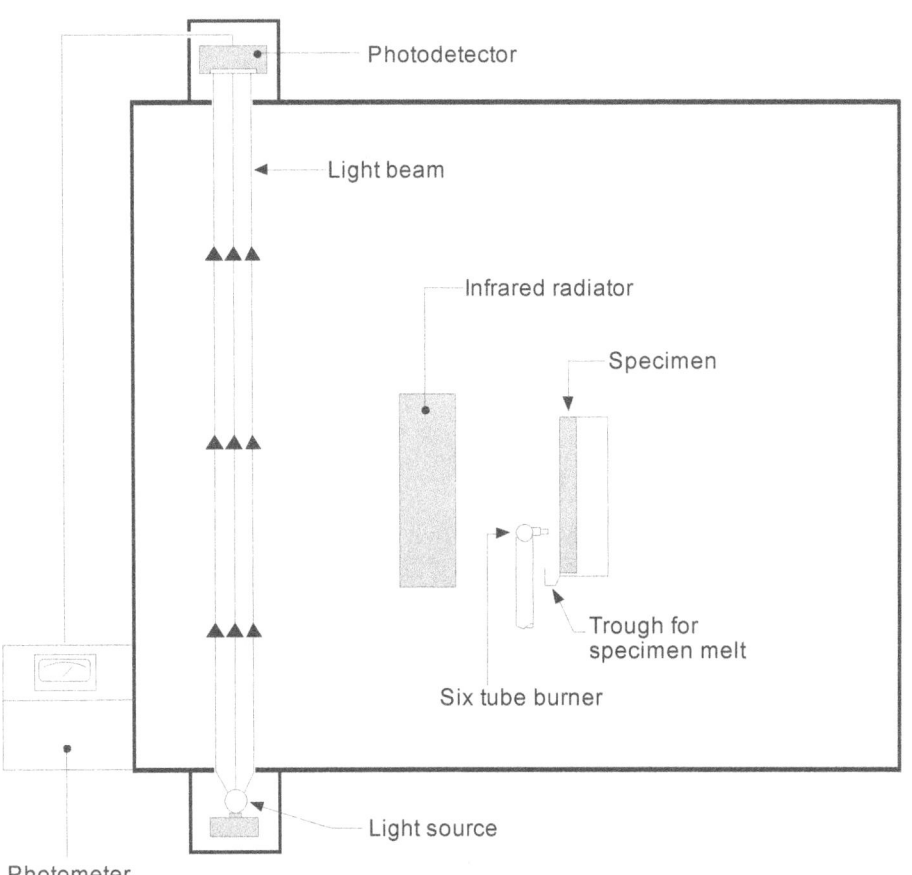

Figure 5. The smoke density chamber, ASTM E 662.

where V is the volume of the chamber (m^3), A is the area of the exposed sample (m^2), L is the path length of the light beam through the smoke (m), I_0 is the intensity of the light beam before the start of the test, and I is the intensity of the light beam during the test. The maximum optical

26

density, D_m, is used primarily in ranking the relative smoke production of a material and in identifying likely sources of severe smoke production. Concerns with the test include the relatively low heat flux exposure, vertical sample mounting, and oxygen supply within the closed test cabinet [38]. Several researchers have concluded that tests like the Smoke Density Chamber do not provide a representation of the smoke emissions to be expected in real-scale fires [39–42]

The International Organization for Standardization Smoke Chamber Test, ISO 5659 (also standardized as ASTM E 1995), shown in Figure 6, uses the same closed test chamber as the ASTM E 662 test, but with an improved specimen holder and exposure system [43,44]. The test apparatus includes a conical heater that can provide a heat flux level as high as 50 kW/m^2 to a 75 mm (3 in) by 75 mm (3 in) horizontal test specimen. Either flaming or non-flaming ignition is supported. Like the ASTM E 662 test, opacity measurements are with a vertically-oriented photometric system. In addition, mass optical density can be obtained with an optional load cell that continuously monitors the mass of the test specimen.

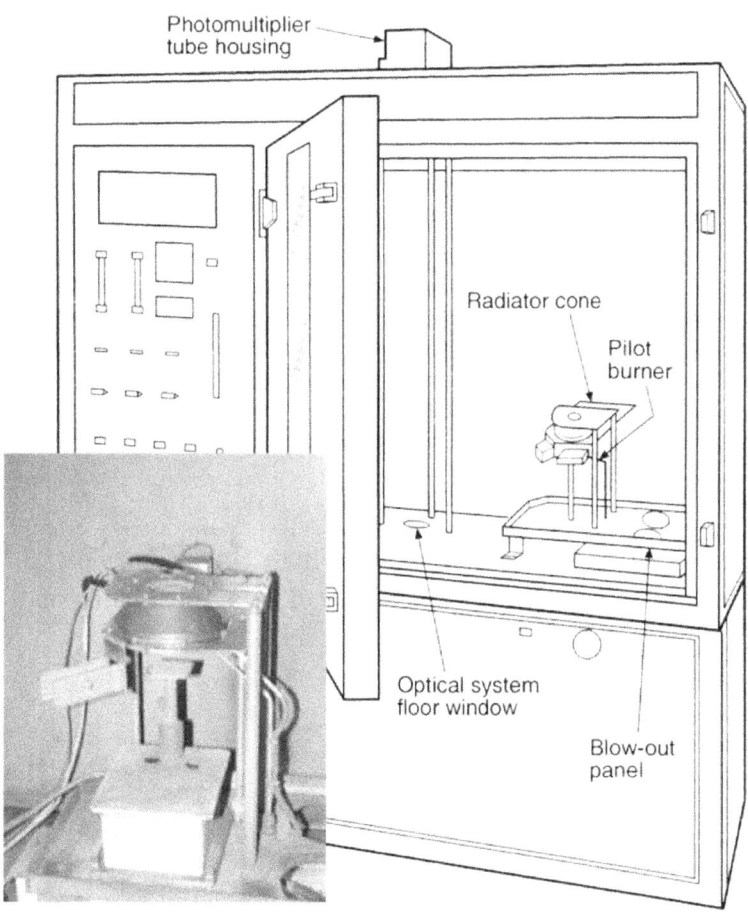

Figure 6. The ISO 5659 smoke density chamber with inset showing conical radiant heater.

While the test retains the limitations of a closed chamber like the ASTM E 662 test, it provides a more uniform sample heating that can be varied over a wider range to provide more realistic exposure. In addition, the load cell permits direct measurement of mass optical density.

Dynamic tests which measure smoke obscuration have typically evolved from existing test methods that measure other fire properties such as heat release rate. These include the Ohio State University (OSU) calorimeter, ASTM E 906 [45], and the cone calorimeter, ASTM E 1354 [46].

The ASTM E 906 apparatus, Figure 7, is used largely for aircraft applications. The test exposes a vertically- or horizontally-oriented specimen to radiant heat. Sample ignition may be non-piloted, by ignition of evolved gases, or by flame impingement on the surface of the sample. The sample is contained within a chamber through which a constant flow of air passes. Measurement of the temperature change and opacity of the gas leaving the chamber allows determination of the rates of both heat and smoke release as the specimen burns. While an upper limit for sample exposure heat flux of 100 kW/m^2 is specified in the standard [45], difficulties are noted with heat fluxes above 50 kW/m^2 [47]. Specimen size is typically 150 mm (5.9 in) by 150 mm (5.9 in), but may be reduced if specimen heat release rate is large enough to allow flames into the hood of the apparatus. For smoke measurement, a photometer is incorporated on top of the exhaust hood. Issues with measurement of heat and smoke release rates have limited the usefulness of the apparatus and most work has transitioned to use of the cone calorimeter.

The cone calorimeter (see Figure 8) makes use of an electric radiant heater in the form of a truncated cone, hence its name. The apparatus is general-purpose in that it may be used to test products for various applications. The heater is capable of being set to a wide variety of heating fluxes from 0 to 100 kW/m^2. The design of the heater was influenced by the ISO test for radiant ignition, ISO 5657 [48]. The technical features are documented in several references [49–52]. Some of the most salient features include:

- horizontal or vertical specimen orientation,
- composite and laminated specimens can be tested,
- continuous mass loss load cell readings,
- feedback-loop controlled heater operation,
- HRR calibration using methane metered with mass flow controller,
- smoke measured with laser-beam photometer and gravimetrically, and
- provision for analyzing CO, CO_2, H_2O, HCl, and other combustion gases.

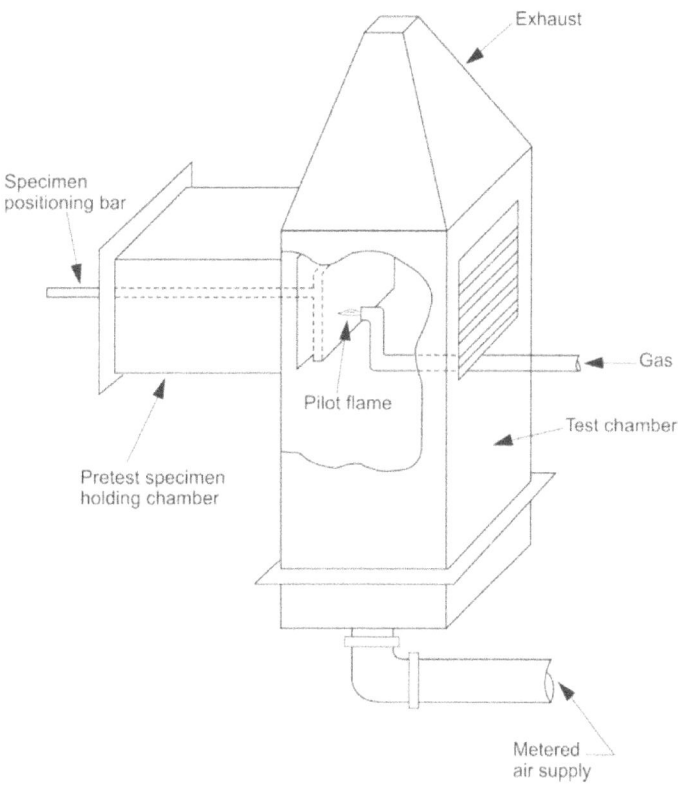

Figure 7. The OSU calorimeter test, ASTM E 906.

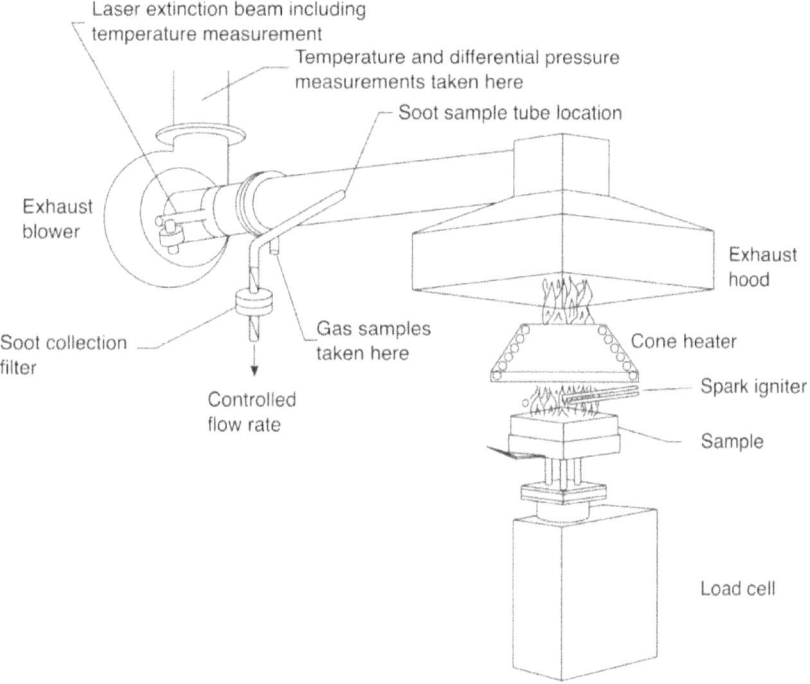

Figure 8. The cone calorimeter, ASTM E 1354.

29

The cone calorimeter is standardized as ASTM E 1354 [46] and ISO 5660 [53]. Smoke obscuration in the cone calorimeter is measured by a laser-based smoke measurement system in the exhaust stream. Smoke is reported from the test as the specific extinction area, the product of the extinction coefficient and volumetric flow rate divided by the mass loss rate.

Hirschler [54,55] and Grayson [38] provide excellent critiques of bench-scale smoke measurement.

3.2   Corrosion Test Methods

A number of test methods have been developed and standardized for testing the corrosive hazard due to smoke. While the study of corrosion due to combustion products is certainly not new [56,57], the effort to develop a modern test method is fairly recent [58]. Initially, testing of the corrosivity of fire effluent was based on indirect measurements such as the amount of halogenated acids the fire effluent produces, the pH, or conductance of the fire effluent. While the tests are relatively straightforward to perform, there is currently no means to take the data from any of these indirect standards and estimate corrosion damage in a real fire. The inherent limits of indirect methods have resulted in the development of a number of direct methods to measure corrosivity of a test material's smoke. Direct methods expose a test probe directly to the smoke and measure the amount of corrosion that occurs.

After the 1974 electronic system fire in Poitiers France, the Centre National d'Études des Télécommunications (CNET) developed what is considered the first modern test of corrosivity [59]. In 1987, ASTM committee E 5.21.70 and the International Electrotechnical Commission (IEC) committee SC 50D each held initial meetings with the objective of developing standard corrosion tests. Initially the ASTM 5.21.70 committee developed ten criteria for an acceptable test for corrosion [60] as follows:

1. Test should measure performance
2. Combustion module should simulate real fire energies and growth rate
3. All products should be tested in the same manner
4. Combustion conditions should be capable of being varied
5. The exposure module should reflect real life conditions
6. The exposure target should be capable of being varied
7. Protocol should allow a reasonable time between exposure and measurement
8. Protocol should consider transport of the combustion products and their decomposition on the target surface
9. Test protocol should not require excessively long use of equipment or operator time
10. Test equipment should be relatively low priced.

IEC SC 50D first considered the matter of corrosion in 1982 but it wasn't until 1987 that IEC SC 50D WG 3 [61] held its first meeting, which in part addressed the issue of the corrosivity of fire effluent. They began preparation of a draft summary of the future guidance document on corrosivity. Subsection titles from their initial report suggest important factors in the test method as follows:

- Relevant fire models
- Relevant corrosion targets
- Effluent transport
- Scaling factor
- Relevant small scale direct tests
- Development of derived (short time, inexpensive, correlatable) test methods for routine quality control
- Damage both in room of fire and remote in space and time to fire
- Misuse of tests and test data

As both of these lists suggest, corrosion is a complex process, and developing an ideal test for the corrosivity of smoke from fires requires broad expertise.

In the early 1990s, a number of tests using direct measures of corrosion were developed. In addition to the CNET test, the cone corrosimeter and the traveling furnace tests were developed. An additional direct test was developed by the ASTM E 05.21.70 committee but it was eventually withdrawn. Because the ASTM E 05.21.70 developed test figures prominently in the literature it will also be discussed in this section. In this section, test methods for corrosion will be described and analyzed.

Recent research concluded that the largest risk to electrical circuits during a fire incident in NPPs is due to bridging of circuits [11]. While this is a serious issue, it has yet to garner much attention in the fire protection community. The last subsection will address how well existing corrosion tests do to predict bridging and what correlations may or may not exist between other fire tests for wire and cable.

3.2.1 Indirect Methods

All the standards for indirect methods of analysis to be described share a basic framework. The effluent from a burning sample is bubbled through some liquid to capture gases. The liquid is then analyzed and the properties reported. The acidity of the effluent is considered important, and because corrosion is an electro-chemical process, the conductivity of the liquid is often also considered. There are two types of indirect tests, one specific to halogenated acids and a second, general type that looks at pH and conductivity. Indirect methods can only address ASTM requirements [60] 1, 2, 9 and 10 directly. While correlations with damage, a strong theoretical understanding, and modeling capability could possibly make indirect methods the preferred test methods, currently such correlations and understanding do not exist. For this reason indirect methods have fallen out of favor as a tool to help address corrosion of fire effluent.

3.2.1.1   Tests for Halogenated Gases: IEC 60754-1 and EN 50267-2-1

Based on experience with real fires, halogenated plastics and halogens in general have been recognized as an important cause of corrosion. For that reason, there are a number of standards that focus on strength of the halogenated acids other than fluoric acid. They do this by using a chemical titration method that is specific to halogens instead of the general pH of a solution.

IEC 60754-1 [62] and EN 50267-2-1 [63], both essentially the same standard, determine the generation rate of halogenated acid produced by a test sample. The standard states that for accuracy this method should not be used for samples containing less than 5 mg/g equivalent of HCl. Values are reported as the equivalent milligrams of HCl produced by 1 g of the sample.

A test sample of 500 mg to 1000 mg is heated in a tube furnace at a uniform rate for 40 minutes until the sample's temperature is 800 °C (1472 °F). It is then kept at that temperature for 20 minutes of the actual test. A constant airflow is supplied to the furnace and then bubbled through two wash bottles of 0.1 M sodium hydroxide solution (NaOH). After the solution has cooled it is titrated using a wet chemistry method specifically meant to determine the amount of halogen acid in the solution and the results are transformed into an equivalent HCl mg/g production rate.

3.2.1.2 Tests for pH and Conductance: IEC 60754-2, EN 50267-2-2 and EN 50267-2-3

The more general pH tests are generally the same as for the halogenated gases except that the analysis is done with a general pH method and an additional test is done to measure the conductance of the solution.

IEC 60754-2 [64], EN 50267-2-2 [65] and EN 50267-2-3 [66] use the same test apparatus as the halogenated gases, but the effluent is bubbled through two wash bottles of 450 mL of deionized water. The furnace is set to at least 935 °C (1715 °F) and run for only 30 minutes with no initial heating phase unlike the halogenated version of the test.

At the conclusion of combustion, IEC 60754-2 and EN 50267-2-2 combine the wash bottles and test the pH and conductivity using appropriate equipment. The values of two replicates are averaged and reported. Annex A of EN 50267-2-2 recommends that if the pass/fail criteria is not specified, the average pH should be greater than 4.3 and the conduction should be less than 10 μS/mm.

For EN 50267-2-3, the reporting is different. Each material component of a cable or other application is tested with three replicates. The mean and standard deviation are determined and if the relative standard deviation is larger than 0.05, an additional three replicates are done and the mean and standard deviation of all six is used. Next the weight per unit of length for each component is measured and the weighted pH is calculated using

$$pH' = \log_{10}\left(\frac{\sum_i w_i}{\sum_i \frac{w_i}{10^{pH_i}}}\right) \tag{3}$$

where w_i is the weight of the ith component and pH_i is the pH for that component. The weighted conductance is calculated using

$$c' = \frac{\sum_i c_i w_i}{\sum_i w_i} \tag{4}$$

where c_i is the conductance for the ith component.

3.2.2   Direct Methods

Direct methods test the effect of fire effluent directly on test samples. The push for direct fire corrosion tests comes from observation that fires can generate corrosive atmospheres even if they do not generate acid gases [67, 68]. While polymers containing halogens and producing acid gases such as HCl or HBr have been shown to cause extensive corrosion damage, many other materials such as woods, wool and polymers that contain nitrogen have also been shown to cause corrosion damage, at times worse than halogenated acids, while not producing halogenated or any other type of acids [69]. Besides measuring the corrosion caused by smoke, a second difference between indirect methods and direct methods is that direct methods can also determine the time dependent aspect of corrosion including the impact that may occur hours, days or even months later.

After the 1974 electronic system fire in Poitiers France, CNET developed what is considered the first modern test of corrosivity [70]. A number of other standards have since been developed that utilize a variety of fire and exposure models as well as several different methods of reporting results. They generally use the same principle for measuring corrosion, which is to use the relationship between the thickness of a conductor and its resistance. The principle for making this measurement will be given and analyzed later.

A common element of all direct tests is a combustion source used to generate the effluent which is exposed to a test probe for a period of time. The test probe is then kept in a controlled environment (for typically 24 h) at a specified temperature and relative humidity. Test measurements usually look specifically at the loss of metal from the test probe.

All the direct tests use the same general method to measure the loss of material due to corrosion. The resistance of a conductor is related to its physical dimensions by

$$R = \rho \frac{l}{A} \tag{5}$$

where R is resistance, ρ is the resistivity of the material, l is the dimension in the direction of the current and A is the cross sectional area of the material. For a rectangular cross section, $A = wh$, where w is the width and h is the height. If the only exposed surface is the top of the conductor, then any change in resistance can be related to the amount of material that is removed by corrosion from the top resulting in an increase in resistance of the conductor.

Direct test methods are broken into two types, static and dynamic. Test methods of each type are described below.

3.2.2.1   Static Methods

Tests are referred to as static tests if all the fire effluent is captured and held during the test. Generally the effluent is captured in the exposure chamber with the target present from the beginning of the test. Capturing all the effluent allows for the target to be exposed for a set period of time that can be significantly longer than the burning time of the sample. Static tests also make it easier to control the ambient conditions of the test such as the temperature of the gas and setting the relative humidity of the test chamber. The test emulates electrical equipment in the upper layer of the room of fire origin.

3.2.2.1.1 ISO 11907-2, Plastics -- Smoke generation -- Determination of the corrosivity of fire effluents -- Part 2: Static method

CNET in France developed what is generally considered to be the first modern corrosion fire test. This test method has been standardized as ISO 11907-2. The basic principle of the test as the standard states is "combustion – condensation – corrosion" [71].

The test chamber is a 20 L (0.7 ft^3) tube with a diameter of about 300 mm (11.8 in). One end has a sample holder where approximately 600 mg of the test sample is ignited and burned. At other end, a water-cooled test target holder keeps the target 10 °C below the chamber temperature to encourage condensation on the test target. The test target is made up of 36 conductor lines with dimensions of 52 mm (2 in) long by 0.3 mm (0.01 in) wide by 17 μm thick on a laminated epoxy circuit board. The nominal resistance, pre-test, is 8.0 Ω ± 0.5 Ω. The reported value for the test is the percentage change of the resistance of the test target. An initial resistance R_i is taken before the test, and at any time t, the resistance is again measured giving R_f. The values are plugged into the following equation, which corrects for variations in initial resistance, to give the test value

$$R_{cor} = 100 \left(\frac{R_i R_f}{8(R_f - R_i)} - 1 \right)^{-1} \tag{6}$$

The derivation of this equation is given by Bottin [72].

Test chamber conditions are set to 50 °C (122 °F) and 65 % relative humidity with the water-cooled test target set to 40 °C (104 °F). The test sample is put in an inert crucible in the sample holder with the heating wire at 800 °C (1472 °F) and kept at that temperature for 3 minutes. After 60 minutes from the time the sample is placed in the sample holder, the final reading of resistance is taken, and the test target is removed from the chamber.

By using the percent change in the resistance as the reported value of the standard, the CNET test has a nonlinear relationship between the reported value and the amount of corrosion that occurs. The CNET categories are at 10 %, 20 % and 30 % change in resistance. Using the above equation, it can be seen that a 10.0 % increase in resistance would mean only 9.1 % of the conductor had been lost to corrosion, a 20.0 % increase would indicate a removal of 16.7 % of the conductor and a 30.0 % increase would mean only 23.1 % of the material had been lost to corrosion.

3.2.2.1.2 ASTM E05.21.71 (withdrawn)

A second static test underwent a significant amount of development in ASTM E05.20. However, after ASTM D5485 was approved, development on ASTM E05.21.71 was discontinued and the standard was withdrawn. Since a great deal of work was done in the development of publications and test data referencing the standard, it is important to describe the test method.

ASTM E05.21.71 starts with the "NIST tox box" as the exposure chamber, which has a volume of 200 L (7 ft^3). It uses a cylindrical quartz combustion cell that is connected to the exposure chamber by a stainless steel lined flue. Exposure ports are used for probe instrumentation and an additional port is connected to a 49 L (1.7 ft^3) plastic expansion bag for pressure relief.

Test sample sizes were not standardized, but the samples were intended to be small enough such that combustion would not be oxygen limited. Test probes were manufactured by Rohrback-Cosasco. The basic configuration was of two legs of copper. One leg is shielded from the effluent and used as at the reference. The other leg is exposed and is the measurement leg.

The test procedure used by Kessel, *et al.* [73] to expose the test sample places the specimen holder in the furnace and the furnace and test chamber are sealed. The atmosphere in the chamber is brought to 60 % relative humidity, the furnace heaters are set to 50 kW/m^2, and the spark igniter is turned on and left running for 15 minutes. After 15 minutes, the spark igniter is turned off and the flue to the test chamber closed. The test probe is then left in the test chamber for another 45 minutes. The probe is then removed from the test chamber and put into an environmental chamber with 75 % relative humidity for 24 hours, and finally, put in to a second environmental chamber with 35 % relative humidity for six days.

Measurements are made every 2 minutes to 3 minutes during the first hour, after 24 hours in the first environmental chamber and after the six days in the second environmental chamber.

3.2.2.2 *Dynamic Methods*

Dynamic methods differ from static methods in that the exposure model has a continuous flow of effluent and does not collect and hold all the effluent for the test duration.

3.2.2.2.1 ISO 11907-3, DIN 53436, Plastics -- Smoke generation -- Determination of the corrosivity of fire effluents -- Part 3: Dynamic decomposition method using a travelling furnace

The traveling furnace test uses an annular furnace that has a 100 mm (3.9 in) exposure area. The furnace is set up using a 1 m quartz glass tube with an outer diameter of 40 mm ± 1 mm and a wall thickness of 2 mm ± 0.5 mm. The test sample is placed in a quartz glass cuvette that is 400 mm ± 10 mm long and 15 mm ± 1 mm high with a wall thickness of 1.7 mm ± 0.2 mm. The test sample preparation differs depending on the type of material but it approximately fills the cuvette.

The test probe is the same as the one used for the CNET tests and like the CNET tests is mounted on a water cooled specimen holder to enhance condensation. It is placed so the center of the probe is 200 mm (7.9 in) from the downstream end of the test sample cuvette.

The test procedure is to place the sample in the cuvette and place the cuvette in the quartz tube so that the first 100 mm (3.9 in) is in the exposure zone of the furnace. Using room air, a flow through the tube of at 100 L/h ± 5 L/h is established. The furnace is set at 600 °C ± 20 °C using the calibration curve previously determined with the reference body. The furnace then travels at 10 mm/min ± 0.5 mm/min until it reaches the other end of the test sample, which should take 30 minutes.

Within 5 minutes of the end of the test the test probe is to be placed in an environmental chamber at a constant 23 °C ± 2 °C and 75 % ± 5 % relative humidity. Resistance measurements are made at one hour and 24 hours after the start of the test and are reported using the same calculation as the CNET tests.

By using a fire model of assisted opposed flow flame spread, the traveling furnace does not model flame spread typical of real fire scenarios, but it is likely a more reproducible fire environment. Ideally, the test probe is exposed to a quasi-steady state flow including products from ignition through to burnout. Having quasi-steady state conditions for most of the 30 minutes of the dynamic part of the test should increase repeatability and reproducibility, which if results can be correlated with real scale fire damage allow meaningful ranking of materials.

3.2.2.2.2 ASTM D5485, ISO 11907-4, Standard Test Method for Determining the Corrosive Effect of Combustion Products Using the Cone Corrosimeter

The cone corrosimeter is designed to be an attachment for the well-established fire performance test, the cone calorimeter. The general idea of the corrosimeter was first published in 1990 by Ryan *et al.* [74].

The corrosimeter attaches to the cone calorimeter above the cone furnace. It starts with a metal funnel above the cone to concentrate the effluent. One end of a rigid, heated, stainless steel tube is placed in the upper narrow opening of the funnel with the opening pointed up away from the burning test specimen so that the opening doesn't become clogged with soot. The other end of the 675 mm ± 75 mm tube is connected to a flexible heat-resistant tube that is 255 mm ± 10 mm long and connects to the exposure chamber. The exposure chamber is 11.2 L ± 0.5 L with an inlet and outlet port and an O-ring sealed top to allow the test probe to be placed in the chamber. The outlet port is connected to sufficient filtering to protect a pump that draws the air through the exposure chamber to be released into a hood or other exhaust handling equipment.

To test a material or configuration of materials requires five samples. The first two samples are run in the cone calorimeter without the corrosimeter to determine the average time to 70 % mass loss. The final three samples are burned with the corrosimeter attached. The flow through the exposure chamber is set at 4.5 L/min. The stainless steel tube is heated to 105 °C (221 °F). The flux from the cone heater is set at the desired value and the material is ignited. Gas from the effluent is drawn until the previously determined 70 % of expected mass loss is achieved. At this point the intake and outlet ports are closed off and the target probe's exposure continues until an hour after the beginning of the test. At this point the target probe is tested to determine the change in resistance, and then the target probe is placed in an environmental chamber with 23 °C ± 2 °C and a relative humidity of 75 % ± 5 % for 24 hours, after which the change in resistance is again measured. To calculate the amount of metal, the standard directs the tester to refer to the procedure provided by the target probe's manufacturer.

While the cone corrosimeter is defined as a dynamic test, it has aspects of both a dynamic test as well as a static test.

3.2.3 Analysis of Individual Standards and Test Methods

Before looking at work done in support of the standards described earlier, it is important to understand what is missing. The most important fact to understand about all the standards is that they only provide a rank ordering of materials. While it is possible to establish classification values on the results of each of the standards, there is currently no work that relates the standards to large-scale testing. Babrauskas and Peacock [75] argue that what makes the cone calorimeter useful for fire protection is that performance in the cone can be related to full scale burns and real fires. The lack of large-scale testing for corrosion implies there is not an equivalent basis for corrosion performance criteria. The best that can be done is to provide a relative ranking of material performance and set performance criteria that would exclude undesirable materials from selection. However, just because a material is relatively noncorrosive in one or all of the bench-scale tests does not necessarily imply that it will exhibit similar properties in a full-scale fire. Thus, there needs to be bench- to large-scale comparisons for the data to be useful for more than relative comparisons.

The Polyolefins Fire Performance Counsel (PFPC) performed a study of four test methods and published the results in five papers [73,76-79]. The first four reviewed the work on each of the test methods, the Radiant Panel, the CNET test, the modified DIN indirect test method and the cone corrosimeter. The final paper discusses and compares the test methods. For completeness, the 25 materials tested by the PFPC authors are included here as Appendix A.

There is little in the literature discussing indirect test methods other than simple test method descriptions. The measurement methods use well-developed wet chemistry methods. However, Saitoh and Inukai [80] used these wet chemistry methods to look at the effect of burning chlorine containing materials on the corrosion of steel and stainless steel plates. Saitoh and Inukai burned 1.0 g samples of various carpet tiles for three minutes in a quartz tube furnace set to either 550 °C (1022 °F) or 750 °C (1382 °F). Airflow of 1 L/min was supplied at one end and the flow of air and effluent was bubbled through an absorbtion bottle of 0.1N NaOH to capture the HCl, and the production rates of HCl were determined.

Using the same furnace and procedures, identical samples were burned, but the furnace was connected to an exposure chamber where plates of steel and stainless steel were mounted in horizontal and vertical orientations. After a 30 minute exposure the plates were stored in an environmental chamber kept at 20 °C (68 °F) and relative humidity of 80 %. After one, two, and four weeks, the amount of corrosion was determined although all corrosion data reported is after four weeks.

As can be seen in Figure 9(a) and Figure 9(b), the corrosion of the steel plate seems to be correlated with the production rate determined for each test sample even though the production rate and corrosion rate were generated from separate experiments; however, the corrosion of the stainless steel plates has no clear correlation with production rates. While the data for the production rate of HCl and the corrosion of the metal were conducted separately, it is clear, at least for the steel plate, that the results of an indirect test can be related to corrosion, but the stainless steel test shows that the correlations don't hold for all materials.

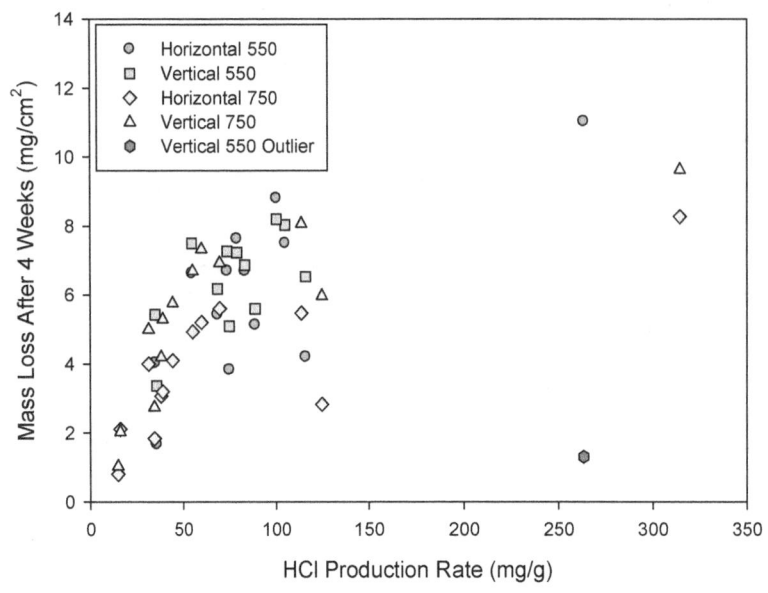

(a) Steel Plate Corrosion

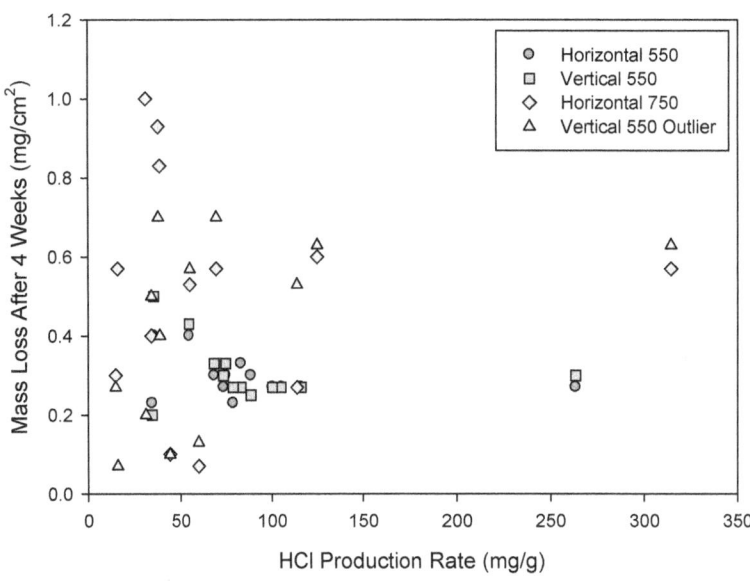

(b) Stainless Steel Corrosion

Figure 9. Comparison of corrosion on plain and stainless steel plates after exposure to fire smoke. Mass loss typically determined by conductive wire resistance change over time [80].

Saitoh and Inukai contend that humidity is probably the most important factor in determining the amount of corrosion that occurs [80]. They also note that the combustion gases from materials such as wood and asphalt that are not expected to produce HCl still produce corrosion. They surmise the presence of other gases besides HCl, which questions the value of test methods that only look for HCl or specific halogen acids.

Both IEC 60754-1 and EN 50267-2-1 do not specifically test for the presence of HF. Sandmann and Widmer [81] looked at the effect of fluoride on stainless steel. They found that the fluoride-containing compounds tested were very stable thermally. They also found that even when the compounds did give off fluoride, the resulting effluent was significantly less corrosive than other halogens like chloride (except for a single test with carbon steel as the target). They noted that HF was significantly less corrosive to copper and silver than HCl was. This seems to contradict Tewarson [19] where HF was found to be significantly more corrosive than HCl. The difference may arise from the target being used for measurement, the method of measurement or the actual environments and possible all three. Tewarson used a steel probe that was put in a solution of the fire effluent in water and measured the corrosion using change in resistance. Perhaps the only conclusion to be drawn is that test conditions can significantly impact observed corrosion.

The CNET test has existed in some form since at least 1983 [82]. It is used as a comparison standard and as such is discussed as it relates to other tests, but there is little work that focused directly on the standard. Rio and O'Neill [59] described the test and presented test data for 12 materials that have corrosion values ranging from near 0 % increase in resistance to close to 15 %. They define the repeatability qualitatively as "reasonable."

The PFPC study used the CNET test as one of its four standards for comparison [76]. They found that the test method did not differentiate the materials particularly well. One modification of the test method that was used was to add 100 g of polyethylene (PE) to the 600 g of test material to ensure complete combustion. Even with the PE, they found that the completeness of combustion and variation in the percent of mass combusted is significant. One possible source for the problem of complete combustion was the three minute heating cycle, which was deemed to be too short. They noted that weight loss seemed to stop for most materials at the end of the three minute period. They also found that for the test probe specified in the CNET test, the risk of bridging between the conductor lines was significant. They showed pictures of two probes after tests that seemed to show some bridging occurring. They also had two tests where the resistance didn't increase but decreased, indicating significant bridging had occurred. Finally, only two of the 24 test materials were found to be statistically more corrosive than the rest.

There is little in the literature about the traveling furnace test used as a smoke corrosivity test. The apparatus has been extensively used and studied in the toxicity test as part of DIN 53436. Prager *et al.* [83] and Prager [84] evaluated the test method in DIN 53436, the traveling furnace, and found that it could properly simulate smoldering and flaming fires. Its utility is demonstrated by its use to collect fire test data on toxicity in developing a significant life-cycle assessment model [85]. Barth *et al.* [86] concluded that the traveling furnace could also be used for corrosion testing. It is seen to have a number of positive attributes including control of combustion for both flaming and smoldering fires. It can generate a relatively constant smoke stream over an extended period or a dynamically changing smoke stream and the composition can be characterized at different times during combustion.

Ryan *et al.* [68] suggested using the cone calorimeter to resolve several problems they found with other test methods they analyzed. First they wanted to address the ASTM E05.21.71 task group's criteria for a bench-scale corrosion test, second, include standard samples of electronic components and circuits, and finally, to include the effect of corrosion on metal loss, circuit bridging, and degradation of electrical contacts. They considered a number of options including using a single chamber for combustion and exposure or dual chambers, using flow-through or closed design and other design parameters. The design they discussed is broadly similar to the standardized cone corrosimeter design, but is significantly different in details. The design purposed used a syringe style extraction method in the duct above the exhaust fan of the cone. The exposure chamber was completely closed. The actual standard as described before has a hybrid open/closed system that draws effluent above the cone radiator, but below the exhaust fan.

The PFPC ran tests using a development version of the cone corrosimeter [78]. The most significant difference between the test method used by PFPC and the one in the standard is the rate of airflow. The stated airflow the PFPC used was 0.024 m^3/s, which is 320 times the 0.000075 m^3/s in the ASTM cone corrosimeter standard and would cause an air change of 0.0112 m^3 in less than 0.5 s. There is no way to determine what airflow rate was actually used from the papers. It is possible they used 0.024 L/s, which would be 0.000024 m^3/s, much closer to the standardized value. In the PFPC tests, both 2500 Å and 45000 Å probes were used and tests were conducted at both 25 kW/m^2 and 50 kW/m^2. The PFPC tests found a number of difficulties with the cone corrosimeter tests. Their first observation was that a brominated polyethylene was found to be less corrosive than three different materials containing fillers. Several samples bubbled out of the sample pan and invalidated the tests, which they felt needed to be corrected. Finally, they found a number of problems with the measurements the test provided.

Most of the difficulties came from using the four separate measurement scenarios recommended in the annex of the standard. The standard doesn't specify a particular heat flux and it doesn't specify a thickness of target but suggests using the 250 nm (2500 Å) probe and/or the 4500 nm (45 000 Å) probe and fluxes of either 25 kW/m^2 or 50 kW/m^2. The PFPC study used both probes together in test series at each flux level and found that there was a significant lack of consistency in the four scenarios. They made six runs with the 6th specimen material, three on one day and three on another day. They found that the repeatability was poor and that, interestingly, the results were strongly correlated by the day of test as can be seen in Figure 10.

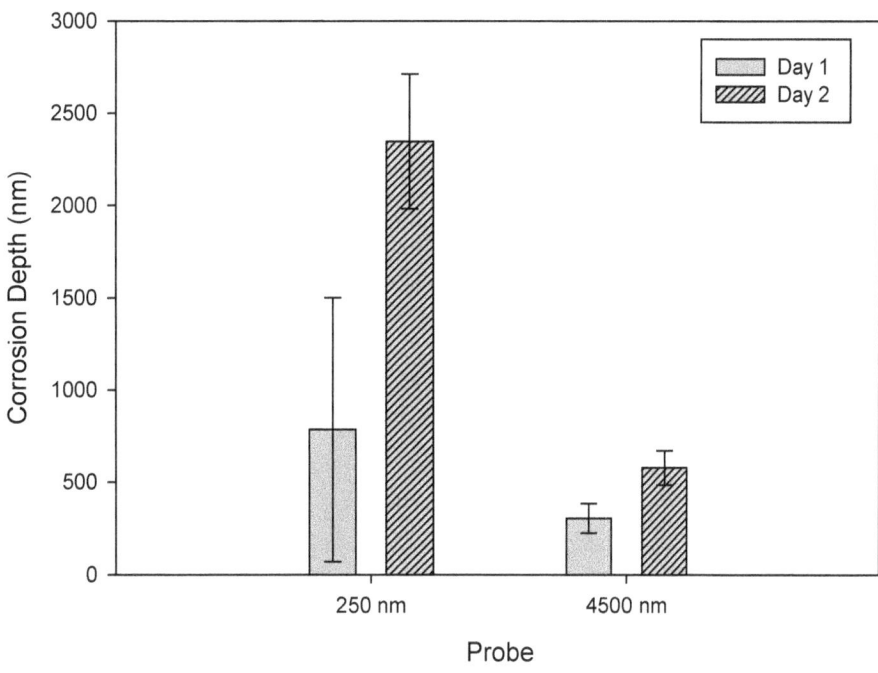

Figure 10. Observed repeatability for a material tested in the cone corrosimeter [77].

The correlation is even stronger than just by the day of test. Because each test had a 250 nm probe and a 4500 nm probe, the test-to-test correlation was strong as seen in Figure 11. The strength of the correlation seems to reinforce the variability in the performance of the material and not just uncertainty in the test measurement. Some possible explanations are offered in Appendix B. There is an issue with pitting that can overestimate the total corrosion of the effluent. This effect is increased with the use of the thinner probe. That would seem to account for the fact that 250 nm probes consistently read a higher level of corrosion. The other issue with the corrosimeter is that the hybrid dynamic/static nature of testing makes it sensitive to differences and variability of HRR between materials and tests. The strong correlation with the day the test was run has no clear explanation. It implies that some variable that was not accounted for changed between the two days, but it is impossible to tell from the paper.

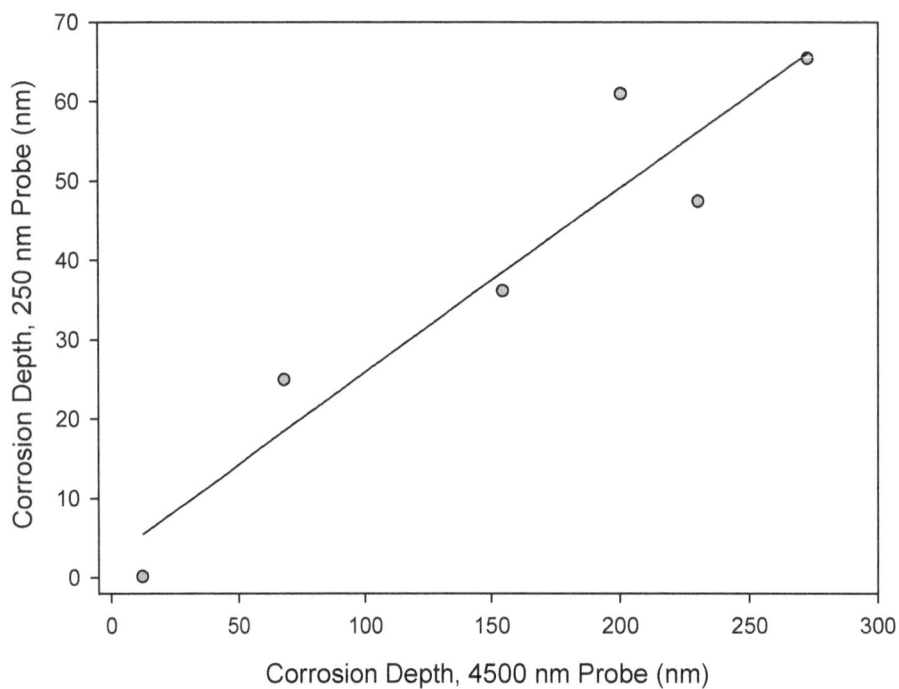

Figure 11. **Comparison of test results in the cone corrosimeter at 25 kW/m^2 and 50 kW/m^2 for one material. Data from reference [77].**

Grand [87] first discussed using a modified version of the NIST cup furnace used for toxicity testing [88, 89] as a test method for both corrosion and toxicity testing. The apparatus is able to determine an ignition time, rate of smoke evolution, total amount of smoke produced, and characterization of the smoke in terms of either toxicity or corrosivity. The test method's flexibility in fire conditions as well as it being a closed system were listed as its key advantages. In 1991, Grand [90] proposed a modified version of the original proposed test focusing on corrosion. A copper probe using resistance to measure metal loss was added to the system. The system was tested with six plastics:

- Ethylene-vinyl acetate (EVA)
- Ethylene-vinyl acetate with fire retardant using red phosphorus (FR-EVA)
- Low density polyethylene (LDPE)
- Low density polyethylene with fire retardant using red phosphorus (FR-6400)
- Polyethylene with fire retardant using bromine (PE165)
- Polyvinyl chloride (PVC).

These tests were run using both 25 kW/m^2 and 50 kW/m^2, irradiating the samples for 15 minutes in each test. The tests generated the expected results with PVC, clearly the worst material in both total metal lost and the rate of metal loss. The PE165 was a distant second and the pure EVA and LDPE were the best. Grand also showed that humidity increased the corrosion rate and total corrosion, as theory and experience predict [87].

Grand also showed time histories of the amount of metal lost in different tests. The utility of this information is clear from the very different shapes of the curves from different test materials. Some materials had a relatively high rate of corrosion that slowly decreased over time while other materials caused short but very rapid loss of material before sharply changing to a much slower rate of loss so that they looked almost like step functions.

Grand summarized the development of the radiant panel test [91]. He argues that the radiant panel test is flexible and useful for characterizing the corrosivity of a material's effluent. He lists five advantages of this system, which include:

- Flexible combustion conditions
- Capturing all fire effluent from the test
- Multiple corrosion probes can be placed in the large test chamber
- The method has proven an ability to distinguish between different polymer materials
- Products can be tested in their end use.

Grand also identified a number of fundamental research issues that needed to be addressed for all corrosion testing and not just the radiant panel. They included:

- The response of resistance probes to hot, moist clean atmospheres
- The repeatability of resistance probe measurements
- The response of resistance probes to known quantities of known corrosive agents
- The impact of orientation and exposure time on measurements
- The importance of using metals besides copper in resistance probes.

The PFPC also studied the radiant furnace test [73]. They found that the test differentiated between different types of polymers. The main concern expressed was the repeatability of the test. For a single material, they had three tests in which metal loss was between 15 Å and 30 Å, two with metal loss between 30 Å and 120 Å and one with metal loss measured at 784 Å. Even excluding the 784 Å case, the standard deviation was very high with a mean of 170 Å ± 303 Å including the 784 Å case, and 47 Å ± 42 Å excluding it.

A gas analysis found that the minimum oxygen concentration was high enough to be assured that combustion was not oxygen limited. Looking at sample size, they found that increasing the mass did not increase the corrosion that occurred by the same factor. They also found that seven out of nine of the smaller samples didn't seem to reach flaming combustion, but weight loss data showed a significant amount of decomposition did occur.

Finally, because some of the test materials were able to corrode the 2500 Å probe beyond its maximum measurement capability, they also used probes of 25,000 Å copper foil. There seemed to be a linear relationship between the thicknesses of probes and corrosivity. They theorized that the copper foil had fewer lattice defects than the dc magnetron sputter-formed 2500 Å probes they used for all other tests. They note as described in Appendix B that increased pitting would significantly change the resistance even though little metal had actually been lost.

3.2.4 Comparison of Test Methods

As mentioned before, Hirschler and Smith [67] showed that all fires are corrosive and that it is not just acid gases that cause the corrosion. In this work that was originally presented in the October 1987 'Corrosive Effects of Combustion Products' conference in London, the results of a number of experiments were shown. The work exposed steel coupons and copper mirrors in an ASTM E662 "NBS Smoke Density Chamber" to the effluent of a number of products for an hour and then weighed the amount of material that was lost to corrosion.

While they did test copper mirrors, the primary focus of the work was the corrosion of steel. They were able to test the corrosivity of the same products in a number of environmental conditions. Corrosion of steel was measured by determining the amount of steel left after the exposure period. Two chamber temperature conditions were used, near ambient, and heated to between 100 °C (212 °F) and 110 °C (230 °F). In the near ambient tests, one coupon was placed on a bag of ice to enhance condensation. For the cases where the chamber was heated, one coupon was placed on a hot plate and warmed to 550 °C (1022 °F). Both test series had three untreated coupons in each test as well as a coupon that was pretreated with machine oil. Two of the three untreated coupons were treated a day later to simulate cleaning. All tests used a high humidity environment.

The results showed that when conditions were varied, different materials tested as the most corrosive. While all materials tested had a corrosive effect on the targets, there was not a consistent pattern in the reaction of the steel coupons to the test conditions. Enhancing condensation with ice typically increased the corrosion though there were exceptions. Heating the coupons was seen to increase corrosion with increased reactivity with oxygen reported as the main reason for the corrosion.

The pretreatment as well as post treatments generally seemed to work, but not always. The treatments, both pre and post, seemed to work better in the heated tests than in the ambient tests. One exception was a single test with a warm exposure chamber where the pretreatment seemed to cause a ten-fold increase in corrosion. All cases were found to render the copper mirrors useless as conductors. At ambient conditions, the materials that released the most HCl caused the most corrosion. While most of the copper in the mirrors remained after the tests, pitting left the mirrors useless as conductors.

Finally, Hirschler and Smith [67] showed that water condensation greatly enhances corrosion. The significant sensitivity of corrosion to test conditions as well as the inherent corrosivity of fire effluent is a significant part of the reason for moving from indirect methods to direct methods.

In 1988, Fallou [61] gave an overview of the subject and a review of activities over the previous ten years that the IEC had conducted on the corrosion issue. According to Fallou, it was generally agreed that it was desirable to be able to use preliminary tests to determine which materials were usable. The hope was that with application of the science of corrosion one would be able to use the results of bench-scale tests to determine full-scale behavior without having to run the experiments. Fallou listed three main concerns with corrosion of electrical equipment.

The first concern was electrical contacts since experience in industrial atmospheres had demonstrated that even low level exposures could cause contacts to fail quickly. The second concern was the loss of metal conductor in circuits causing the circuit to fail. Thirdly, impact of corrosives on insulators which can lead to bridging or short circuits was a concern.

In 1988, Briggs [92] reported on the status of the actual tests under development for three properties of smoke, density, toxicity and corrosivity. His main focus was on smoke density and smoke toxicity. He reports that the interests of the ISO/TC61/SC4/WG2 group were mainly in understanding the decomposition models, condensation techniques and corrosion detectors. At that time, the fire tests being considered were a Japanese system, DP5659, for measuring smoke density and the traveling furnace in DIN 53436. No specifics were offered on the work on exposure chambers. Finally, initial work on a corrosion probe would use the CNET probe.

Hirschler gave a summary of the state of understanding of smoke corrosivity at the 1990 International Wire & Cable Symposium [93]. He compared the results of two acid gas tests, the hot tube furnace used by the Canandian Standards Association and the "coil" test with the CNET probe. He noted that the acid gas tests give virtually identical results and that the CNET results "parallels" the acid gas results. He suggested that this is because the CNET test forced condensation, with no post exposure period and an unrealistically intense fire model. While the CNET results have the same trend as the acid gas tests, the results do not parallel the acid gas test results.

Hirschler reviewed his and Smith's work [67] described earlier, putting significant focus on the impact of heat on corrosion. He notes how heating the steel coupon to 550 °C (1022 °F) in a moist environment with the air heated to 100 °C (212 °F) to 110 °C (230 °F) causes a significant and relatively consistent level of corrosion. While the results are significant, it is not clear how realistic the scenario actually is. Hirschler notes that chlorine-generating materials are more corrosive in the cooler environment, possibly due to condensation. He also points out that nylon's increased corrosivity in the warmer environment didn't have an obvious explanation.

As the first of the direct test methods, the CNET test is a benchmark used for comparison of other tests. Bottin showed that the CNET method was an appropriate method to measure corrosivity of fire effluent [72, 94]. As part of the testing, Bottin compared the results of the CNET test to the conductivity of seven samples that were tested. The two samples containing fluorine were not tested for conductivity, but for the other five samples, only one product failed to rank the same for both tests, a flame retarded compound containing decabromodiphenyloxide. As can be seen in Figure 12, the four materials that did rank the same in both tests seem to be almost perfectly exponentially correlated.

In 1991, Briggs [69] reviewed IEC 754-1 and IEC 754-2, which were standards at the time, as well as the traveling furnace and radiant panel tests, which were both in a development phase. The traveling furnace became IEC 754-3. The cone corrosimeter, which became IEC 754-4, is not reviewed. He discussed the concerns with corrosion testing at the time, which have largely remained unchanged, but didn't review any data, old or new. He concluded that adequate testing methods had not yet been developed.

45

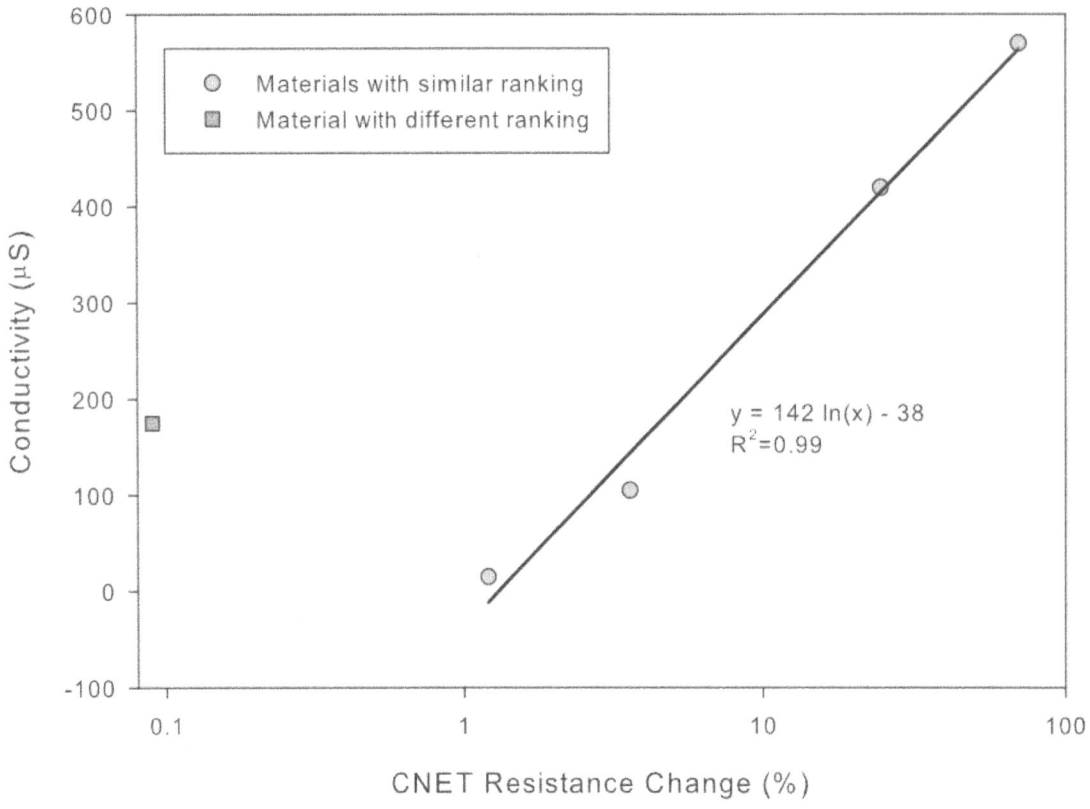

Figure 12. **Comparison of CNET test results with measurements of conductivity for five samples tested. Data from references [72,94].**

Hirschler studied the available test data for smoke corrosion tests [95 - 97]. He described the CNET test, ASTM E05.21.71 draft radiant panel test, ASTM D09.21.4 cone corrosimeter in draft form and the DIN 53463 traveling furnace. He included all the tests that he reviewed in more detail in International Wire & Cable Symposium paper [93]. Added to this paper is a review of the ASTM E05.21.71 task force requirements with recommendations for two additional requirements[4]:

11. Test should be repeatable and reproducible
12. Test results should correlate validly with those obtained from corrosion following full scale fires.

He later looked at the four corrosion test methods of the time and considered how they performed compared to the requirements. He found the indirect tests do not meet most of the criteria. Hirschler thought that most of the problems with the CNET test can be easily resolved but the main problem is repeatability. Basically, there is not much information on the other tests. He reported that the DIN traveling furnace test was not defined well enough to evaluate although

---

4 See page 30 for the original 10 recommended requirements for a corrosion test method from the ASTM E05.21.71 task force.

he stated, "the combustion module of this test is, however, generally considered as unrepresentative of real fire conditions, meaning criteria 2 would not be met". He deemed the radiant panel and cone corrosimeter tests to meet criteria 1 to 7, but neither address criteria 8, 11 and 12.

Gandhi [98,99] performed some simple modeling of three types of gas collections systems that could be used in corrosion tests, the CNET test, the cone corrosimeter, and a third system that utilized a piston cylinder chamber. He used the mass fraction of an arbitrary corrosive component of the fire effluent, a, as an indicator of how each capture system works and developed design parameters for each test. However, the important quantity for determining corrosivity is the concentration. The nondimensionalized concentration measures for the three tests are also developed in Appendix C. For the CNET test, the concentration is much easier than the mass fraction and is simply

$$\frac{dC_a^*}{d\tau} = y_a \Gamma_f^* \tag{7}$$

where C_a^* is the nondimensionalized concentration, τ is the nondimensionlized time, y_a is the production rate of product a in units of kg of product a produced per kg of fuel burned and Γ_f^* is a nondimensionlized mass loss rate. Thus, at the end of the burning phase of the CNET test, the concentration of product a is dependent on the amount of product a produced by the material being tested. With one caveat, there is a very straightforward relationship between the test material and the concentration of a particular corrosive product. The caveat is that there is concern about the mass loss rate being limited by the available oxygen.

To show how serious the concern about oxygen limited combustion is, assume that the oxygen lower limit is 15.5 %, then the maximum sample size that can undergo complete combustion given the stoichiometric oxygen to fuel ratio is calculable as shown in Figure 13. Note that the nondimensionalized mass of a test material in ISO 11907-2 is 1.0. A number of common materials could experience oxygen limited burning in the CNET test so some care must be taken with interpreting the results of the test.

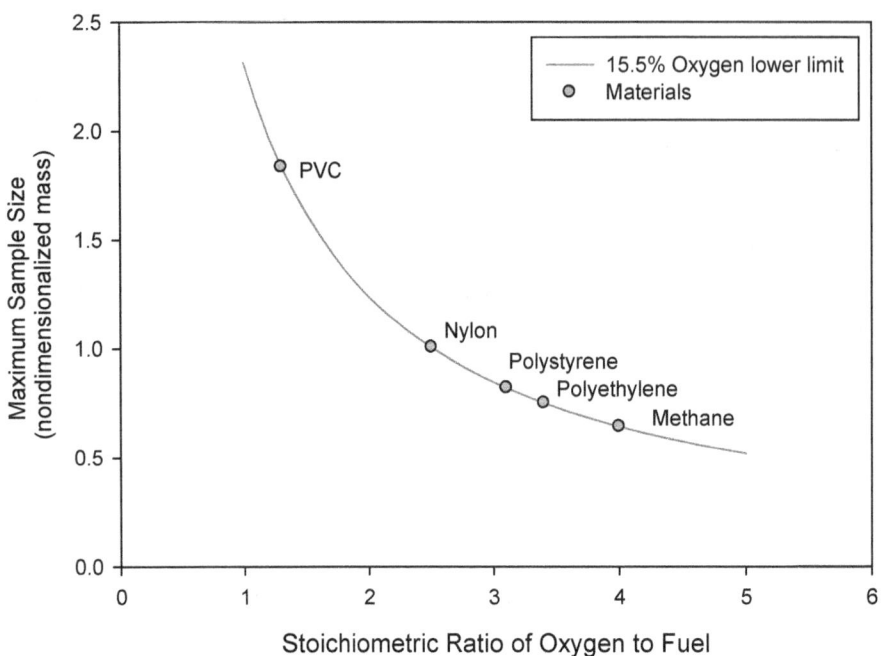

Figure 13.   Maximum sample size for given oxygen fuel ratio assuming complete combustion.

Modeling the cone corrisometer Gandhi makes two significant simplifications. The first is to assume that the test uses mass flow control instead of volume flow control and that the control is sufficiently good to maintain constant mass in the exposure chamber. The second is not so much a simplification as a potential for misunderstanding. Gandhi assumes that the amount of mass entering and leaving is related to the mass loss rate (MLR) by a fraction f. While there is no requirement for f to be a constant it is easy to make that mistake. To make the analysis easier, it is desirable to consider a constant mass flow (see details in Appendix C). Since the volume is constant and the assumption is the mass is also constant, then the mass fraction of product a, Y_a, is also the nondimensionalized concentration and is given by

$$\frac{dC_a^*}{d\tau} = f_{in}^* \left(\frac{y_a}{1 - \alpha} - Y_a \right) \tag{8}$$

where $f_{in}{}^*$ is the nondimensionalized mass flowing into and out of the exposure chamber, Y_a is the fraction of the mass in the exposure chamber that is product a and α is the rate of the mass of air entrained by the fire to the MLR. The term

$$\frac{y_a}{1 - \alpha} \tag{9}$$

represents the fraction of the total flow of mass, both effluent and entrained air, generated by the test sample in the cone. This term will overestimate the fraction of the mass flowing into the

48

exposure chamber at points where the total mass flow from the product is less than the forced flow of the pump, but should not be a significant error.

The fifth paper by the PFPC compared the results of the four standards and performed extensive analysis [79]. They reviewed the entire project and then did an overall analysis of the results. They looked at three factors in the comparison and analysis:

1. Precision – Is the test repeatable? Does the test differentiate corrosive potentials?
2. Accuracy – Does the test differentiate corrosive potential consistent with known chemistry? Are the test results consistent with existing standards?
3. Cost, availability and convenience – Is the test equipment accessible, easy to operate and mechanically sound?

To assess repeatability, they looked at the average coefficient of variation (the standard deviation divided by the mean expressed as a percentage), as well as the standard deviation of the coefficient of variation. Their results are given in Table 8.

Table 8. Repeatability of several corrosion standards. Data from reference [79].

Test		Average Coefficient of Variation	Standard Deviation of Coefficient of Variation
ASTM E05.21.71		86.0	42.3
CNET		51.8	27.1
DIN	Conductivity	11.4	8.95
	pH	5.21	4.65
ASTM	50 kW/m^2 45 000 A	37.0	31.2
D09.21.04	50 kW/m^2 2500 A	22.2	22.8

Clearly the indirect tests were the most repeatable of all the tests. None of the direct tests had a particularly low average coefficient of variation. The question becomes are they sufficiently repeatable. To answer this question they used two statistical tests to see how many statistically differentiated groups each test could break the 24 products into. CNET differentiated product 20 into one group, product 19 into a second group and the rest into a final group.

Overall, they found that none of the four test methods satisfied all the criteria for a useful corrosion test. Since the tests did not give the same rankings they were not interchangeable. They recommended that both the cone corrosimeter and the radiant panel needed to have their repeatability improved and that development on the CNET test should stop.

4 MODELING OF SMOKE GENERATION, TRANSPORT, AND DEPOSITION

4.1 Use of Smoke and Species Yields as Model Inputs

While there are detailed combustion models available that fully address gas-phase chemistry, they are not practical for the large domains encountered in typical fire scenarios. As such, modeling of chemistry in most current computer fire models is not considered in detail beyond a one-step effective reaction and is largely a bookkeeping exercise based on a simple reaction of fuel and oxygen such as shown below [100].

$$C_v H_w O_x N_y Cl_z + v_{O_2} O_2$$
$$\rightarrow v_{CO_2} CO_2 + v_{H_2O} H2O + v_{CO} CO + v_S \text{Soot} + v_{N_2} N2 + v_{HCl} HCl + v_{HCN} HCN$$

For complete combustion of the simplest hydrocarbon fuel, methane reacts with oxygen to form carbon dioxide and water. The only input required is the pyrolysis rate and the heat of combustion. For fuels that contain oxygen, nitrogen, or chlorine, the reaction becomes more complex. In this case, the user specifies model inputs for the composition of the fuel and any non-ideal combustion yields (such as the production of soot or carbon monoxide) to allow the model to calculate the mass of species produced during combustion. Most typically, these are specified as molar yields (i.e., the number of moles of a given species produced *per* mole of fuel consumed in the combustion reaction), though other normalized values are also used[5]. For example, for soot, the effective molar yield, v_S, is related to the soot yield, y_S, via the relation

$$v_S = \frac{W_f}{W_S} y_S \tag{10}$$

where W_f and W_S are the effective molar mass of the fuel and soot, respectively. In most cases these are idealized estimates for actual complex fuels.

The total amount of a given species released by a fire is thus dependent on the fuel consumed. This is most typically specified as a user input based on a combination of the burning object's heat release rate ($\dot{Q}$), mass loss rate ($\dot{m}$), and heat of combustion (H_c) related as $\dot{Q} = \dot{m} H_c$. The discussion above is essentially book-keeping, the accounting of the gas molecules formed in the combustion process using estimates or empirical values for v_S, v_{HCl}, etc. The heat release rate is also dependent on the amount of oxygen and local temperature; that is combustion will only take place if there is sufficient oxygen and temperature for the reaction. Thus, unburned fuel may also be transported from the fire source and perhaps combusted in regions removed from the initial fire location. The level of detail in local burning is model dependent[6].

---

[5] For example, the CFAST model [100] specifies combustion for most species relative to the production of carbon dioxide; the input for production of carbon monoxide is the mass of carbon monoxide produced relative to the mass of carbon dioxide produced.

[6] In CFD fire models, combustion takes place within a single grid cell (control volume) dependent on the local conditions within the cell. In the simpler zone fire models, combustion is determined by the oxygen available in a

4.2   Smoke and Species Transport

Transport of mass and energy in computer fire models, including individual species, is simply based on a mass and energy balance at the boundaries. The level of detail naturally depends on the complexity of the model. For zone models, species transport, including soot, is tracked as it is generated by a fire, transports through vents from compartment to compartment, or mixed between layers (control volumes) within a compartment. For computational fluid dynamics (CFD) fire models, the transport is tracked from individual grid cells (control volumes) to adjacent cells.

In either case, each unit mass of a species produced by a fire is carried in the flow to the control volumes and may accumulate within these. The model keeps track of the mass of each species in each volume, and knowing the volume of each control volume, one can easily calculate the concentration of each tracked species as a function of time. Detailed chemistry effects are not included, but are represented as a global representation.

4.3   Modeling of Deposition on Compartment Surfaces

Smoke deposition on compartment surfaces is an important phenomenon that needs to be understood to assess the effects of smoke far removed from the fire source. Both smoke particulates and gas phase species will deposit or stick to and react with compartment surfaces providing a sink to the fluid mass balance. A fairly accurate accounting of the smoke concentration at the location of critical electrical equipment vulnerable to failure is necessary to make a quantitative assessment.

For hydrogen chloride, it has been shown [101,102] that significant amounts of the substance may be removed through adsorption by surfaces that contact smoke. HCl production is treated in a manner similar to other species. However, an additional term is required to allow for deposition on, and subsequent absorption into, material surfaces. Limited empirical data are available for deposition of HCl on several typical compartment surfaces that provide correlations for the adsorption and absorption of HCl. Data for other surfaces or acid gases is not available.

FDS [103] includes a model for smoke deposition on bounding surfaces [104]. It is based on simplified thermophoresis and turbulent deposition mechanisms that exclude particle size effects. It has not been fully validated, but appears to improve the smoke concentration prediction somewhat [105].

4.4   Modeling of Equipment Damage

Equipment damage from smoke stems primarily from an increased resistance in circuits and connections by corrosive metal loss, and creating conductive paths for current leakage due to smoke deposition. The former may take hours or days after the fire is extinguished to manifest

---

fire plume (in turn based on empirical correlations) or in a larger control volume within a compartment. Thus, oxygen-limited combustion is naturally a far more detailed process in a CFD model than in a zone fire model.

itself in equipment failure, while the latter may occur at failure levels any time soon after initial smoke exposure.

Though more typically based on laboratory measurement that compares the relative corrosion resistance of two or more materials, simple models for estimating the expected service life of components based on these laboratory measurements have been proposed [106,107]. Corrosion rates are expressed as an infinite series with the effect of environmental factors taken individually and in interaction. In many cases only a relatively small number of terms of this series would typically be required to make a prediction within the inherent uncertainty of the system. The laboratory measurements then look at the expected environment and the sensitivity of the rate in the expected mean environment to variations in each parameter to determine the coefficient for each term in the series. While the structure of the model has been defined, limited experimental data has been developed to date to support such a model.

Tanaka *et al.* [6] have proposed a regression model to predict the degradation of surface insulation resistance based on smoke exposure tests. The regression model is not generic; it only applies to the fixed exposure time in the specific test apparatus described. It cannot be implemented in a fire model in a predictive manner where smoke concentrations, temperatures, humidity and flow fields are changing with time. Mangs, and Keski-Rahkonen [108] describe a model to predict insulation resistance deterioration based on the amount of smoke deposited, and ohmic conduction through the deposited soot. Hagen *et al.* [109] describe soot deposition on a probe consisting of parallel gold traces. The resistance behavior observed was non-ohmic and exhibited a critical loading consistent with percolation theory.

Gallucci [110] has proposed a simplistic model to estimate the probability of damage to electronic devices based on potentially synergistic effects of exposure to smoke particles and/or elevated humidity. With data on smoke deposition from Tanaka *et al.* [6] and on probability of damage to electrical devices from Karydas [111], he developed a simple relationship between the probability, p, of damage and smoke concentration of

$$p = 1 - \frac{1}{e^{CT-3/16}} \tag{11}$$

where T is the smoke obscuration (m^{-1}) and C is a constant taken to be 0.038 for vertical surfaces and 0.202 for horizontal surfaces. The effects of humidity are seen to accelerate the failure. The acceleration factor can be estimated from the temperature and humidity. A simple relationship is suggested as

$$k = 10^{\frac{\theta+RH-165}{40}} \tag{12}$$

where θ is the temperature (°C) and RH is the humidity (%). The value of k is limited to the range of 1 to 1000. While its recommended use is for sensitivity or bounding analyses, it does include the important parameters of soot deposition, elevated temperature, and elevated humidity. All are seen as areas in need of additional experimentation and modeling.

5 FUTURE RESEARCH NEEDS

5.1   Smoke Production and Transport

Butler and Mulholland give a very good synopsis of the generation and transport of smoke components [112]. They present the current state of knowledge about smoke aerosol phenomena that affect smoke toxicity: soot generation, fractal structure of soot, agglomerate transport via thermophoresis, sedimentation, diffusion, agglomerate growth through coagulation and condensation, and the potential of the aerosols to transport adsorbed or absorbed toxic gases or vapors into the lungs. The phenomena that affect smoke toxicity do or may play a role in the non-thermal damage of equipment due to smoke exposure except transport of absorbed or absorbed toxic gases or vapors into the lungs. The analog to transport of absorbed or adsorbed toxic gases or vapors into the lungs is transport of absorbed or adsorbed corrosive gases or vapors onto electrical conductors, so their discussions on absorption and adsorption are relevant here. They include tables of measured smoke yields and aerodynamic particle sizes, equations and references for the smoke agglomerate transport properties and wall loss.

They conclude that the quality of fire hazard and risk assessment with regard to toxicants in smoke would be improved by conducting research in the following areas:

- Measurement of mass median aerodynamic diameter of soot agglomerates avoiding possible agglomerate structural changes with impactors.
- Quantitative information on the adsorption of irritant gases on fire generated soot aerosol.
- Quantitative information on the losses of toxicants to walls for a range of realistic fires.
- Development of model for predicting smoke aerosol and vapor loss to the walls for a fire in an enclosure.
- Information on the size distribution of water droplets at fires, the conditions under which they are formed, and the amount of gases adsorbed on the droplets.
- Understanding the role of nanoparticles on the toxic effect of perfluoropolymer fumes.

All but the last item would help facilitate or improve smoke transport fire modeling and the consequences of smoke exposures far from the fire room.

The level of detail needed to describe smoke properties of interest depends on the effect of the smoke one wants to estimate. Light extinction can be estimated from the smoke concentration and the specific extinction coefficient [113] while particle size distribution, fractal structure, and optical properties are needed for predicting light scattering from soot [114]. It is not clear what the important smoke properties are, and the concentrations, or deposition conditions that significantly affect electrical or electronic equipment failure.

At the moment, smoke generation is an empirical input for modeling purposes, with simple transport models and deposition models. Actually, smoke generation is a function of ventilation and thermal conditions for even pure materials. For complex fuels, combustion chemistry, gas chemistry, and liquid chemistry all contribute to products that arrive on and may react with electrical and electronic equipment.

5.2   Modeling Smoke Deposition

FDS [103] includes a simple model for smoke deposition to provide a gross estimate of smoke particle losses. It lacks detailed physics to describe particle mass loading on surfaces under specific conditions including gravitational settling. Typical forces acting on particles that may drive them to surfaces include diffusional forces (both Brownian motion and eddy diffusion), thermophoresis, inertial forces including gravitational settling, and electrical forces experienced by charged particles in electric fields and induced electrostatic forces between particles and non-conducting surfaces. In general, the dry deposition velocity characterizes the smoke mass deposition per unit area via the equation

$$V_{ds} = \frac{\text{Flux}}{\text{Concentration}} \quad\quad\quad (13)$$

where V_{ds} is the dry deposition velocity (m/s), Flux is mass flux to the surface (g/m^2·s), and Concentration is the mass concentration (g/m^3). Thus, given a smoke mass concentration in the air next to a surface, the flux is the product of the concentration times the dry deposition velocity. Unfortunately, the dry deposition velocity is a function of all the forces acting on particles of a given size. Models have been proposed to estimate the dry deposition velocity of spherical particles of known size on semiconductor wafers [115,116]. Particles sized less than 0.1 μm in diameter fall into the diffusion-controlled deposition regime, where the deposition velocity increases with decreasing particle size. For vertically orientated surfaces, or horizontal face-down surfaces, the deposition velocity will continuously decrease as the particle size increases. For horizontal face-up surfaces, a minimum in the dry deposition velocity is observed at about 0.2 μm to 0.3 μm, and then it rises due to gravitational sedimentation, yielding a valley-shaped curve. A thermal gradient caused by a cooler surface than the surrounding fluid will increase the dry deposition velocity for all particle sizes, while a thermal gradient due to a hotter surface will decrease the dry deposition velocity for all particle sizes.

In principle, if all details of particle size, and charge, air and surface temperature, flow fields, electric fields, and surface orientation are known, the dry deposition velocity as a function of particle size could be computed. If the aerosol concentration and size distribution are known, then the total flux to the surface could be calculated from summation of the contributions from each particle size. Typically though, the particle size distribution is not properly tracked in fire models, nor is there a detailed model of dry deposition velocity of agglomerate particles based on a representative agglomerate size.

5.3   Prediction of Equipment Damage

It is speculated that the main failure mechanisms for electrical and electronic equipment from smoke exposure during a fire in a NPP are due to circuit bridging from smoke deposits [6,25].

There is no absolute correspondence between the change in surface insulation resistance from smoke exposure and the failure of any electronic or electrical components. A relationship between surface insulation resistance from smoke deposition and the failure of real electronic circuits needs to be studied to establish correlations that could be used in probabilistic risk assessments. A smoke damage routine developed for a fire model could then assess the near term (during and soon after exposure) damage potential of electronics and electrical components due to smoke exposure. Fire models could provide inputs to a smoke damage estimates which would most likely include temperatures, humidity, flow fields, orientation, smoke concentration and properties. The smoke damage estimates could account the electronics or electrical component orientation, surface temperature, voltage potentials, and identified susceptibilities. The complexity and uncertainty and simplification of model inputs suggest a sub-model based on physics and empirical correlations for the predicted effects and the probability of equipment failure.

6 CONCLUSIONS

This report presents a review of smoke production measurement, prediction of smoke impact as part of computer-based fire modeling, and measurement and prediction of the impact of smoke through deposition of soot on and corrosion of electrical equipment. Equipment damage from smoke stems primarily from an increased resistance in circuits and connections by corrosive metal loss, and creating conductive paths for current leakage due to smoke deposition. The former may take hours or days after the fire is extinguished to manifest itself in equipment failure, while the latter may occur at failure levels any time soon after initial smoke exposure.

The literature review on smoke corrosivity testing and damage due to smoke deposition emphasizes (despite extensive research on smoke corrositity) the lack of validated and widely applicable prescriptive or performance-based methods to assure electrical equipment survivability given exposure to fire smoke. Corrosion failures following smoke exposure take days or longer to manifest themselves, and corrective actions in the interim time can take place if the exposure is not too severe and intervention is timely.

Circuit bridging via current leakage through deposited smoke was identified as an important mechanism of electronic and electrical equipment failure during NPP fires. In order to assess the potential for damage due to smoke exposure, relationships between smoke exposure and the failure of real electronic components need to be established. A smoke damage routine developed for a fire model could then assess the near term (during and soon after exposure) damage potential of electronics and electrical components design fire exposures.

In the longer term, better understanding of combustion, gas-phase, and liquid chemistry could allow estimates of the formation rate of smoke and acid gases as a function of the ventilation, thermal, and fire environments. Understanding of absorption of acid gases by smoke and its deposition on surfaces (as a function of temperature, humidity, and smoke reactivity) would be necessary to develop models that could estimate the deposition behavior of smoke.

7 REFERENCES

1 "Cable Fire at Browns Ferry Nuclear Power Station," NRC Bulletin BL-75-04, U.S. Nuclear Regulatory Commission, Washington, DC, March 1975.

2 Jacobus, M. J. "Screening Tests of Representative Nuclear Power Plant Components Exposed to Secondary Environments Created by Fires," NUREG/CR-4596 and SAND86-0394, U.S. Nuclear Regulatory Commission, Washington, DC, June 1986.

3 "Program for Sampling, Analysis, and Cleanup of Residue on Affected Structures, Systems, and Components," Brown's Ferry Nuclear Plant, April 12, 1975, Unpublished.

4 "Recommendations Related to Browns Ferry Fire," NUREG-0050, U.S. Nuclear Regulatory Commission, Washington, DC, February 1976.

5 "Hinsdale Central Office Fire," Joint Report of Office of the State Fire Marshall and Illinois Commerce Commission Staff, March 1989.

6 Tanaka, T. J., S. P. Nowlen, and D. J. Anderson, "Circuit Bridging of Components by Smoke," Sandia National Laboratory, Albuquerque, NM and U.S. Nuclear Regulatory Commission, Office of Nuclear Regulatory Research, Washington DC: SAND96-2633 and NUREG/CR-6476, 1996.

7 "Fire Effects on Electrical and Electronic Equipment," Hughes Associates, Inc., in DOE Handbook Fire Protection, Volume II, DOE-HDBK-1062-96, 1996.

8 Allan, R. "The Great New York Telephone Fire," IEEE Spectrum, June, 1975.

9 Wiggins, J. H., "Economics Consequences of the Hinsdale, Illinois Bell Fire of May 8, 1988," NSF/ENG-89043, National Science Foundation, Washington, DC, September 1989.

10 Patton, J. S., "Fire and Smoke Corrosivity of Metals," *J. of Fire Sciences*, 9:149-161, March/April 1991.

11 EPRI/NRC-RES "Fire PRA Methodology for Nuclear Power Facilities: Volume 2: Detailed Methodology." Electric Power Research Institute (EPRI), Palo Alto, CA, and U.S. Nuclear Regulatory Commission, Office of Nuclear Regulatory Research (RES), Rockville, MD: EPRI TR-1011989 and NUREG/CR-6850, 2005.

12 "NFPA 70, National Electrical Code," 2008 Edition. National Fire Protection Assn., Quincy, MA (2008).

13 Kuligowski, E. D., "Compilation of Data on the Sublethal Effect of Fire Effluent." National Institute of Standards and Technology, Technical Note 1644, 2009.

14 Wanless, J., "Investigation of the Potential Fire-Related Damage to Safety-Related Equipment in Nuclear Power Plants." Sandia National Laboratory, Albuquerque, NM and U.S. Nuclear Regulatory Commission, Office of Nuclear Regulatory Research, Washington DC: SAND85-7247 and NUREG/CR-4310, 1985.

15 Jacobus, M., "Screening Tests of Representative Nuclear Power Plant Components Exposed to Secondary Environments Created by Fires." Sandia National Laboratory, Albuquerque, NM and U.S. Nuclear Regulatory Commission, Office of Nuclear Regulatory Research, Washington DC: SAND86-0394 and NUREG/CR-4596, 1986.

16 "383-2003 – IEEE Standard for Qualifying Class 1E Electrical Cables and Field Splices for Nuclear Power Generating Stations," IEEE Standards Association, Piscataway, NJ, 2003.

17 Korsah, K., T. J. Tanaka, T. L. Wilson, and R. T. Wood, "Environmental Testing of an Experimental Digital Safety Channel." Oak Ridge National Laboratory, Oak Ridge, TN and U.S. Nuclear Regulatory Commission, Office of Nuclear Regulatory Research, Washington DC: ORNL/TM-13122 and NUREG/CR-6406, 1996.

18 Tanaka, T. J., "Effects of Smoke on Functional Circuits." Sandia National Laboratory, Albuquerque, NM and U.S. Nuclear Regulatory Commission, Office of Nuclear Regulatory Research, Washington DC: SAND97-2544 and NUREG/CR-6543, 1996.

19 Tanaka, T. J. and S. P. Nowlen, "Results and Insights on the Impact of Smoke on Digital Instrumentation and Control." Sandia National Laboratory, Albuquerque, NM and U.S. Nuclear Regulatory Commission, Office of Nuclear Regulatory Research, Washington DC: SAND99-1320 and NUREG/CR-6597, 2001.

20 Tewarson, A. and F. Chu, "Non-Thermal Fire Damage Due to Smoke and Corrosive Compounds," Factory Mutual Research Corporation, Norwood, Massachusetts, 1991.

21 Reagor, B.T., "Smoke Corrosivity: Generation, Impact, Detection, and Protection," *J. of Fire Sciences*, 10:169-179, March/April 1992.

22 Tewarson, A., "Nonthermal Fire Damage," *J. of Fire Sciences*, 10:188-242, May/June 1992.

23 Tewarson, A., " Characterization of fire environments in central offices of telecommunications industry," *Fire and Materials* 27:131-149, 2003.

24 Tewarson, A, F. Chu and J. P. Hill, "Quantification of Fire Characteristics for Assessment of Nonthermal Fire Damage," Factory Mutual Research Corporation, Norwood, Massachusetts 1992.

25 Chapin, J.T., P. Gandhi, and L. M. Caudill, "Comparison of Communications LAN Cable Smoke Corrosivity", Fire Risk & Hazard Research Assessment Research Application Symposium, San Francisco, CA, June 25-27, 1997.

26 Steiner, A.J., "Method of Fire-Hazard Classification of Building Materials," *ASTM Bulletin*. pp. 19-22, 1943.

27 Babrauskas, V.; R. D. Peacock, E. Braun, R. W. Bukowski, and W. W. Jones. "Fire Performance of Wire and Cable: Reaction-to-Fire Tests--A Critical Review of the Existing Methods and of New Concepts." National Institute of Standards and Technology, NIST TN 1291; 130 p. 1991.

28 Tewarson, A. "Thermal and Nonthermal Damage in Fires," in *International Conference on Fire Research and Engineering (ICFRE). Proceedings.* Lund, D. P.; Angell, E. A., Editors, September 10-15, 1995, Orlando, FL, National Institute of Standards and Technology (NIST) and Society of Fire Protection Engineers (SFPE). SFPE, Boston, MA, 455-460 pp, 1995.

29 Tanaka, T. J., S. J. Martin, and K. M. Hays, "Determination of Critical Shunt Resistance for Circuit Bridging Failures in Logic Circuits." Appendix to [18].

30 "Federal Trade Commission Complaint on the Flammability of Plastic Products," File No. 732-3040 (May 31, 1973).

31 Emmons, H.W. "Fire and Fire Protection," *Scientific American*, 231, 21-27 (July 1974).

32 Östman, B., and R. Nussbaum. National Standard Fire Tests in Small-scale Compared with the Full-scale ISO Room Test (Rapport I 870217), Träteknik Centrum, Stockholm (1987).

33 "ASTM Fire Standards and Related Technical Material – 7th Edition," ASTM International, West Conshohocken, PA (2007).

34 Jin, T., "Visibility Through Fire Smoke," Report of Fire Research Institute of Japan, 2(33), pp. 12-18 (1971).

35 Jin, T., "Visibility through Fire Smoke in Main Reports on Production, Movement and Control of Smoke in Buildings," Japanese Association of Fire Science and Engineering, pp. 100-153 (1974).

36 Jin, T., "Visibility Through Fire Smoke," *J. of Fire and Flammability,* 9(2) pp. 135-155 (1978).

37 "Standard Test Method for Specific Optical Density of Smoke Generated by Solid Materials," ASTM E 662-09, Annual Book of ASTM Standards, Volume 04.07, ASTM International, West Conshohocken, PA (2009).

38 Grayson, S. J. ,"Smoke Opacity Test Methods," in *Hazards of Combustion Products – Toxicity, Opacity, Corrosivity, and Heat Release,* November 10-11, 2008, 171-183, Interscience Communications Ltd., London (2008).

39 Breden, L. H. and M. Meisters., "The Effects of Sample Orientation in the Smoke Density Chamber," *Journal of Fire and Flammability*, 7, 234-247 (1976).

40 Babrauskas, V., "Applications of Predictive Smoke Measurement," *Journal of Fire and Flammability*, 12, 51-64 (1981).

41 Quintiere, J. G. "Smoke Measurements: An Assessment of Correlations Between Laboratory and Full-Scale Experiments," *Fire and Materials*, 6, Nos. 3/4, 145-160 (1982).

42 Babrauskas, V., "Smoke Measurement Results from the Cone Calorimeter," in *Proceedings of the 8th Joint Panel Meeting UJNR Panel on Fire Research and Safety*, May 13-21, 1985, 420-434 (1985).

43 "Plastics – Smoke generation – Part 2: Determination of optical density by a single-chamber test," ISO 5659-2:2006, International Organization for Standardization, Geneva (2006).

44 "Standard Test Method for Measurement of Smoke Obscuration Using a Concial Radiant Source in a Single Closed Chamber, With the Test Specimen Oriented Horizontally," ASTM E 1995-08, ASTM International, West Conshohocken, PA (2009).

45 "Standard Test Method for Heat and Visible Smoke Release Rates for Materials and Products Using a Thermopile Method," ASTM E 906-07, ASTM International West Conshohocken, PA (2009).

46 "Standard Test Method for Heat and Visible Smoke Release Rates for Materials and Products Using an Oxygen Consumption Calorimeter," ASTM E 1354-09, ASTM International West Conshohocken, PA (2009).

47 Babrauskas, V. and S. J. Grayson, "Heat Release Test Methods," ," in *Hazards of Combustion Products – Toxicity, Opacity, Corrosivity, and Heat Release,* November 10-11, 2008, 171-183, Interscience Communications Ltd., London (2008).

48 "Fire tests — Reaction to fire — Ignitability of building products," ISO 5657, International Organization for Standardization, Geneva (1997).

49 Babrauskas, V. "Development of the Cone Calorimeter — A Bench-scale Heat Release Rate Apparatus Based on Oxygen Consumption." *Fire and Materials* 8 (1984).

50 Babrauskas, V., and W.J. Parker. "Ignitability Measurements with the Cone Calorimeter." *Fire and Materials* 11 (1987).

51 Babrauskas, V., and G. Mulholland. "Smoke and Soot Data Determinations in the Cone Calorimeter," in *Mathematical Modeling of Fires*, 83-104, (ASTM STP 983), ASTM International, West Conshohocken, PA (1987).

52 Babrauskas, V. "The Cone Calorimeter — Fire Properties," in *New Technology to Reduce Fire Losses & Costs*, 78-87, S.J. Grayson and D.A. Smith, eds. London: Elsevier Applied Science Publishers (1986).

53 "Reaction-to-fire tests -- Heat release, smoke production and mass loss rate -- Part 2: Smoke production rate (dynamic measurement)", ISO 5660-2:2002, International Organization for Standardization, Geneva (1997).

54 Hirschler, M. M. "How to Measure Smoke Obscuration in a Manner Relevant to Fire Hazard Assessment: Use of Heat Release Calorimetry Test Equipment," *Journal of Fire Science*, 9, 183-222 (1991).

55 Hirschler, M. M. "Key to Measure Smoke Obscuration Measurements relevant to Fire Hazard: Heat Release Rate Calorimetry Test Equipment," in *Fire Safety Developments and Testing: Toxicity – Heat Release – Product Development – Combustion Corrosivity*, October 21-24 1990, Ponte Verde Beach, FL, 127-155, Fire Retardant Chemicals Association, Lancaster, PA, (1990).

56 Benner, R. C., "Chemistry of Soot and the Corrosive Products of Combustion," in *Papers on the Effect of Smoke on Building Materials*, Smoke Investigation Bulletin No. 6, Mellon Institute of Industrial Research and School of Specific Industries, University of Pittsburgh, Pittsburgh, PA, 1913.

57 Benner, R. C., "Effect of Smoke on Metals," in *Papers on the Effect of Smoke on Building Materials*, Smoke Investigation Bulletin No. 6, Mellon Institute of Industrial Research and School of Specific Industries, University of Pittsburgh, Pittsburgh, PA, 1913.

58 Grayson, S. J., "Corrosion Test Methods," in *Hazards of Combustion Products – Toxicity, Opacity, Corrosivity, and Heat Release*, November 10-11, 2008, 171-183, Interscience Communications Ltd., London (2008).

59 Rio, P. and T. J. O'Neill, "Assessing the Risk of Corrosion Damage from Wire and Cable Systems in Fires," in *Cables and Belting International Conference*, September 23-24, 1986, Lancaster, UK, 22/1-8, Plastics and Rubber Institute, Centre National d'Études des Télécommunications

60 Roux, H. J., Title of article, Corrosive Effects of Combustion Products, ASTM, IEC and ISO, London, October 1987. OMC, Fire and Materials Ltd. London.

61 Fallou, B., "IEC Activity Related to Corrosivity and Other Effects of Fire Effluents," *IEC ACOS Workshop II*, April 11-12, 1988, R9001958

62 IEC 60754-1:1994 Test on gases evolved during combustion of materials from cables-Part 1: Determination of amount of halogen acid gases

63 EN 50267-2-1:1998, Common test methods for cables under fire conditions – Test on gases evolved during combustion of materials from cables – Part 2-1: Procedures – Determination of amount of halogen acid gas

64 IEC 60754-2 Amend 1:1997 Test on gases evolved during combustion of materials from cables-Part 2: Determination of degree of acidity of gases evolved during the combustion of materials taken from electric cables by measuring pH and conductivity

65 EN 50267-2-2:1998, Common test methods for cables under fire conditions – Test on gases evolved during combustion of materials from cables – Part 2-2: Procedures – Determination of the degree of acidity of gases for cables by measuring pH and conductivity

66 EN 50267-2-3:1998, , Common test methods for cables under fire conditions – Test on gases evolved during combustion of materials from cables – Part 2-3: Procedures – Determination of the degree of acidity of gases for cables by determination of the weighted average of pH and conductivity

67 Hirschler, M. M., Smith, G. F., "Corrosive Effects of Smoke on Metal Surfaces," *Fire Safety J.*, 15:57-93 Nov 1989. R8901143

68 Ryan, J. D., Babrauskas, V., O'Neill, T. J., Hirschler, M. M., "Performance Testing of the Corrosivity of Smoke," *Characterization and Toxicity, ASTM STP 1082*, H. K. Hasegawa, Ed., American Society for Testing and Materials, Philidelphia, 1990, pp 75-88.

69 Briggs, P. J., "Corrosivity of Fire Effluents: International Developments in Performance-Related Test Procedures," in *Recent Advances in Flame Retardancy of Polymeric Materials: Materials, Applications, Industry Developments, Markets.* Volume 2. May 14-16, 1991, Stamford, CT, Business Communications Co., Inc., Norwalk, CT, Lewin, M.; Kirshenbaum, G. S., Editors, 121-130 pp, 1991

70 Rio, P., O'Neill, T. J., "Assessing the Risk of Corrosion Damage from Wire and Cable Systems in Fires," Plastics and Rubber Institute. Fire Resistant Hose, Cables and Belting. International Conference. September 23-24, 1986, Lancaster, UK, 22/1-8 pp, 1986.

71 ISO 11907-2:1995, Plastics – Smoke generation – Determination of the corrosivity of fire effluents – Part 2: Static method.

72 Bottin, M. F., "Acidity and Corrosivity Measurements of Fire Effluent," *Proceedings of 39th International Wire and Cable Symposium*, International Wire and Cable Symposium, Nov 13-15, Reno, NV, 205-213 pp 1990.

73 Kessel, S. L., Bennett, J. G. Jr., Rogers, C. E., "Corrosivity Test Methods for Polymeric Materials. Part 1 – Radiant Furnace Test Method," *J. of Fire Sciences* 12:109-133, Mar/Arp 1994.

74 Ryan, J. D., Babrauskas, V., O'Neill, T. J., and Hirschler, M. M., "Performamce Testing for the Corrosivity of Smoke," *Characterization and Toxicity of Smoke, ASTMSTP 1082*, H. K. Hasegawa, Ed., American Society for Testing and Materials, Philadelphia, 1990, pp. 75-88.

75 Babrauskas, V. and R. D. Peacock, "Heat Release Rate: The Most Important Variable in Fire Hazard," *Fire Safety J.,* Vol. 18, No. 3, 255-272, 1992.

76 Rogers, C. E., Bennett, J. G. Jr., Kessel, S. L., "Corrosivity Test Methods for Polymeric Materials, Part 2-CNET Test Method," *J. of Fire Sciences*, 12:134-154, Mar/Apr 1994.

77 Bennett, J. G. Jr., Kessel, S. L., Rogers, C. E., "Corrosivity Test Methods for Polymeric Materials. Part 3-Modified DIN Test Method," *J. of Fire Sciences*, 12:155-174, Mar/Apr 1994.

78 Bennett, J. G. Jr., Kessel, S. L., Rogers, C. E., "Corrosivity Test Methods for Polymeric Materials. Part 4-Cone Corrosimeter Test Method," *J, of Fire Sciences*, 12:175-195, Mar/Apr 1994.

79 Kessel, S. L., Rogers, C. E., Bennett, J. G. Jr., "Corrosivity Test Methods for Polymeric Materials. Part 5-A Comparison of Four Test Methods," *J. of Fire Sciences*, 12:196-233, Mar/Apr 1994.

80 Saitoh, F. and Inukai, T., "Corrosivity Test of Smoke from Cl-Contained Materials." U.S./Japan Government Cooperative Program on Natural Resources (UJNR). Fire

Research and Safety. 12th Joint Panel Meeting. October 27-November 2, 1992, Tsukuba, Japan, Building Research Inst., Ibaraki, Japan Fire Research Inst., Tokyo, Japan, 280-288 pp, 1992.

81 Sandmann, H. and Widmer, G., "The Corrosiveness of Fluoride-containing Fire Gases on Selected Steels," *Fire and Materials*, 10:11-19 Mar 1986.

82 Fallow, B., "IEC Activity Related to Corrosivity and Other Effects of Fire Effluents," *IEC ACOS Workshop II*, April 11-12 1988.

83 Prager, F. H., Einbrody, H. J., Hupfeld, J., Muller, B. and Sand H., "Risk Oriented Evaluation of Fire Gas Toxicity Based on Laboratory Scale Experiments-The DIN 53436 Method," *Journal of Fire Sciences*, 5:308-325 Sept/Oct 1987.

84 Prager, F. H., "Assessment of Fire Model DIN 53436," *Journal of Fire Sciences*, 6:3-24 Jan/Feb 1988.

85 Simonson, M., Andersson, P., Rosell, L., Emanuelsson, V., Stripple, H., "Fire-LCA Model: Cables Case Study: SP REPORT 2001:22," SP Swedish National Testing and Research Institute, 2001.

86 Barth, E., Muller, B., Prager, F. H., Wittbecker, F.-W., "Corrosive Effects of Smoke: Decomposition with the DIN Tube According to DIN 53436," *Journal of Fire Sciences* 10:432-454 1992.

87 Grand, A. F.,"Development of a Laboratory Radiant Combustion Apparatus for Smoke Toxicity and Smoke Corrosivity Studies," *11th Joint Meeting of the UJNR Panel on Fire Research and Safety*, Oct 19-24, 1989.

88 Alexeeff, G. F., Packham, S. C., "Use of a Radiant Furnace Fire Model to Evaluate Acute Toxicity of Smoke," *Journal of Fire Sciences*, 2:306-320, Jul/Aug 1984.

89 Levin, B. C., Fowell, A. J., Birky, M. M., Paabo, M., Stale, A., Malek, D., "Further Development of a Test Method for the Assessment of the Acute Inhalation Toxicity of Combustion Products," National Bureau of Standards (U.S.), NBSIR82-2532 1982.

90 Grand A. F.,"Evaluation of the Corrosivity of Smoke from Fire Retarded Products," *J. of Fire Sciences*, 9:44-59, Jan/Feb 1991.

91 Grand, A. F., "Evaluation of the Corrosivity of Smoke Using a Laboratory Radiant Combustion/Exposure Apparatus," *J. of Fire Sciences* 10:72-93, Jan/Feb 1992.

92 Briggs, P. J., "Status Report on ISO Tests for Evaluation of Smoke Density, Corrosivity and Toxicity of Fire Effluents," *IEC ACOS Workshop II*, April 11-12, 1988.

93 Hirschler, M. M., "Update on Smoke Corrosivity," *International Wire & Cable Symposium Proceedings*. Nov 13-15 1990.

94 Bottin, M.-F., "The ISO Static Test Method for Measuring Smoke Corrosivity," *J. of Fire Sciences*, 10:160-169, Mar/Apr 1992. R9200959

95 Hirschler, M. M., "Analysis of Test Results from a Variety of Smoke Corrosivity Test Methods," *Eighteenth International Conference on Fire Safety*. San Francisco CA, Jan 11-15, 1993.

96 Hirschler, M. M., "Discussion of Smoke Corrosivity Test Methods: Analysis of Existing Tests and Their Results," *Fire and Materials*, 17:231-246, 1993.

97 Grayson, S. J., Hirschler, M. M., "Testing for Corrosivity of Smoke in Materials and Products," *1st Fire and Materials International Conference*, Arlington VA Sept 24-25 1992.

98 Gandhi, P. D., "Modeling Gas Collection Systems for Corrosion Testing," *Fire Safety Journal*, 21:47-68 1993.

99 Gandhi, P. D., "Corrosion from Combustion Products, An Overview," *Fire Research and Safety, 13th Joint Panel Meeting*, Gaithersburg, MD March 13-20 1997.

100 Jones, W. W., Peacock, R. D., Forney, G. P., and Reneke, P. A., "CFAST – Consolidated Model of Fire Growth and Smoke Transport (Version 6), Technical Reference Guide," National Institute of Standards and Technology, Gaithersburg, MD, Special Publication 1026 2009.

101 Galloway, F. M. and M. M. Hirschler, "A Model for the Spontaneous Removal of Airborne Hydrogen Chloride by Common Surfaces," *Fire Safety Journal*, 14:251 1989.

102 Galloway, F. M. and M. M. Hirschler, "Transport and Decay of Hydrogen Chloride: Use of a Model to Predict Hydrogen Chloride Concentrations in Fires Involving a Room-Corridor-Room Arrangement," *Fire Safety Journal*, 16:33 1990.

103 McGrattan, K., Hostika, S., Floyd, J., Baum, H., Mell, W., and McDermott, R., "Fire Dynamics Simulator (Version 5) Technical Reference Guide Volume 1: Mathematical Model," National Institute of Standards and Technology, Gaithersburg, MD, Special Publication 1018-5 2011.

104 McDermott, R., McGrattan, K., Hostikka, S., and Floyd, J, "Fire Dynamics Simulator (Version 5) Technical Reference Guide Volume 2: Verification," National Institute of Standards and Technology, Gaithersburg, MD 2010.

105 Cohan, B.D., "Verification and Validation of the Soot Deposition Model in Fire Dynamics Simulator," M.S. Thesis, University of Maryland, College Park, 2010.

106 Ricker, R. E., "Can Corrosion Testing Make the Transition from Comparison to Prediction?," J. of Mater., 47:9, pp. 32-35 1995.

107 Ricker, R. E., "On Using Laboratory Measurements to Predict Corrosion Service Lives for Engineering Applications," J. of ASTM International, 5:7, pp. 1-10, ASTM International, West Conshohocken, PA 2008.

108 Mangs, J., and Keski-Rahkonen, O., "Acute Effects of Smoke from Fires on Performance of Control Electronics in NPP's," Transactions, SMiRT 16, Washington, DE, August 2001.

109 Hagen, G., Feistkorn, C., Wiegartner, S., Heinrich, A., Bruggemann, D., and Moos, R., "Conductometric Soot Sensor for Automotive Exhausts: Initial Studies," Sensors, 10, 1589-1598, doi:10.3390/s100301589, 2010.

110 Gallucci, R. "Smoke-Induced Failure, with Potential Effect of Temperature and Humidity, of Electrical Devices in Fire Probabilistic Risk Assessment," PSAM10 - Tenth International Probabilistic Safety Assessment and Management Conference, Session 15-5, Fire: Risk and Consequence Analysis 2, June 7-11, 2010, Seattle, Washington 2010.

111 Karydas, D., "A Probabilistic Methodology for the Fire and Smoke Hazard Analysis of Electronic Equipment," INTERFLAM '93, London, UK 1993.

112 Butler, K. M., and Mulholland, G. M., "Generation and Transport of Smoke Components" in Fire Technology volume 40. pp. 149–176. 2004

113 Mulholland, G. W.; Croarkin, C., " Specific Extinction Coefficient of Flame Generated Smoke." Fire and Materials, Vol. 24, No. 5, 227-230, September/October 2000

114 Dobbins, R. A.; Mulholland, G. W., and Bryner, N. P., "Comparison of a Fractal Smoke Optics Model with Light Extinction Measurements," Atmospheric Environment, Vol. 28, No. 5, 889-897, 1994.

115 Liu, Y.H., and Kang-ho, A., "Particle Deposition on Semiconductor Wafers," Aerosol Science and Technology, 6:215-224, 1987.

116 Oh, M.D., Yoo, K. H., and Myong, H.K., "Numerical Analysis of Particle Deposition onto Horizontal Freestanding Wafer Surfaces Heated or Cooled," Aerosol Science and Technology, 25:141-156, 1996.

APPENDIX A   MATERIALS INCLUDED IN THE PFPC STUDY

Table A-1. 25 Materials Included in the PFPC Study from Reference [72].

PFPC No.	Company	Product Designation	Material Description
1	BP	EXP 839	XL olelin elstomer with metal hydrate filler
2	Dow	5435-30-11	Blend of HDPE and chlorinated PE elastomer
3	Dow	5348-40-1	Chlorinated PE with fillers
4	Exxon	EX-FR-100	EVA polyofefin with ATH filler
5	GEP	NORYL PX1766	Polyphenylene oxide/polystyrene blend
6	GEP	ULTEM 1000	Polyetherimide
7	GEP	SILTEM STM1500	Polyethermide/siloxane copolymer
8	Himont	Astryn AA36H	Intumescent polypropylene
9	Lonza		Nylon with mineral filler
10	UCC	UNIGARD RE DFDA-1736NT	XL polyolefin copolymer with mineral filler
11	UCC	UNIGARD RE HFDA-1393BK	
12	Quantum	Petrothene XL 7403	
13	Quantum	Petrothene YR 19535	
14	Quantum	Petrothene YR 19543	
15	UCC	UCARSIL FR-7920NT	
16	BP	Polycure 798	
17		Commercial sample	
18		Commercial sample	
19		Commercial sample	
20		Commercial sample	
21	UCC	DGDK-3364NT	
22			Douglas fir
23	Quantum	Ultrathene UE 631	
24	Nylet	P50	
25	UCC	UNIGARD HP HFDA-6522NT	

APPENDIX B EFFECT OF CORROSION PROBE THICKNESS

In Part 4 [78] of the PFPC study, it is noted that for the cone corrosimeter that the amount of corrosion lost as measured by the 2500 Å probe was always higher than that measured by the 45 000 Å probe for the same material. Since corrosion is a localized phenomenon, it is possible for different parts of a conductor to have different amounts of material removed. Considering the test probe as a series of resisters with different heights (due to local pitting) but the same width and length, the total resistance of the probe can be quite different than that of a probe of average thickness throughout.

To show the impact of the thickness of the probe and the local variability of the corrosion, consider Figure B-1. It shows a range of probes with thicknesses from 2500 Å to 40 000 Å exposed to the same test environment. The "average loss" numbers are from the original data and are determined by calculation from total probe mass loss assuming mass is lost uniformly across the probes.

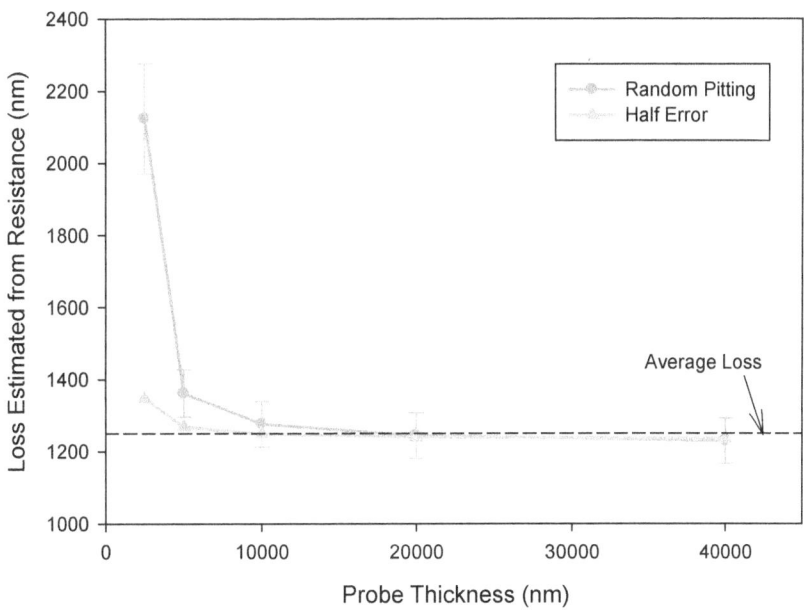

Figure B-1. Modeled measures of metal loss (indicated by resistance change) with pitting.

To estimate the impact of random pitting of a probe, a hypothetical probe was divided into a number of sections with the same width and height, but with a random variation in section height chosen such that the total mass lost was consistent with the average value. As can be seen the thicker the probe, the closer the average of the experiments matches the average mass loss calculation. For thinner probes, this results in individual sections particularly thin with higher resistance values. When the whole probe is considered as resistances in series, this results in a higher overall resistance.

APPENDIX C ANALYSIS OF GAS COLLECTION IN A CORROSION TEST APPARATUS

To extend the type of analysis of Gandhi [97,98] to consider the withdrawn ASTM E05 test, this section provides an idealized analysis of the gas volume within an arbitrary test chamber. The difficulty modeling the ASTM E05 test is the expansion bag, because that leads to a changing volume. Assuming an idealized expansion bag that keeps the pressure in the chamber constant leaves three state variables to describe the chamber: mass in chamber, M, the volume of the chamber V and the temperature in the chamber T. Because the chamber captures and holds all the effluent the change in mass can be described by

$$\frac{dm}{dt} = (1+\alpha)\dot{m}_b \tag{C-1}$$

where m_b is the rate of the mass of the sample burning and α is the ratio of mass entrainment rate to mass burned rate as a function of height above the sample and the HRR of the sample, $\alpha = \dot{m}_{en}/\dot{m}_b = h_c C \eta^m - 1$. An energy balance can be used to calculate the temperature

$$\frac{dE}{dt} = c_p(1+\alpha)\dot{m}_b T_{in} = \frac{d}{dt}(c_p mT) = c_p\left(\frac{dm}{dt}T + m\frac{dT}{dt}\right) \tag{C-2}$$

where E is the enthalpy in the chamber, c_p is the heat capacity of air at constant pressure and T_{in} is the temperature of the mass entering the chamber from the burning sample. Solving for rate of change of temperature yields

$$\frac{dT}{dt} = \frac{(1+\alpha)\dot{m}_b}{m}(T_{in} - T) \tag{C-3}$$

Differentiating the volume using the ideal gas law gives

$$\frac{dV}{dt} = \frac{d}{dt}\left(\frac{m\hat{R}T}{P_0}\right) = \frac{\hat{R}}{P_0}\left(\frac{dm}{dt}T + m\frac{dT}{dt}\right) \tag{C-4}$$

where P_0 is the constant pressure in the chamber and $\hat{R} = R/M_{ave}$ with R being the ideal gas constant and M_{ave} is the average mass of a molecule of air. Notice that the rate of change in the volume is directly proportionally to the rate of change in the enthalpy in the chamber so that

$$\frac{dV}{dt} = \frac{\hat{R}}{P_0}(1+\alpha)\dot{m}_b T_{in} \tag{C-5}$$

Calculating P_0 in terms of initial conditions, making the substitution in the previous equation and grouping terms results in

$$\frac{dV}{dt} = \frac{\hat{R}}{\frac{m_0 \hat{R} T_0}{V_0}} (1+\alpha) \dot{m}_b T_{in} = V_0 (1+\alpha) \frac{\dot{m}_b}{m_0} \frac{T_{in}}{T_0} \qquad \text{(C-6)}$$

which interestingly enough is virtually the nondimensionalized differential equation for the volume.

The three state equations that govern the gas collection system have now been defined. To understand the impact of a gas collection system on the corrosion test, it is important to look at factors that control corrosion, in this case the products produced by burning a test sample.

Gandhi's analysis focuses on the mass fraction, Y_a, of a species a given that some fraction, y_a, of the mass burned becomes a. While this assumption ignores details of any reaction chemistry, it is appropriate for a simple analysis of these gas collection systems. For the ASTM E05, test, consider what happens if a constant fraction of the mass burned, m_b goes to producing a. That produces the following equation for the mass of a in the chamber, which can be written as $M \cdot Y_s$ with Y_s being the mass fraction of a

$$\frac{d(mY_a)}{dt} = \frac{dm}{dt} Y_a + m \frac{dY_a}{dt} = \dot{m}_b y_a \qquad \text{(C-7)}$$

Solving for dY_a/dt gives

$$\frac{d(Y_a)}{dt} = \frac{(1+\alpha)\dot{m}_b}{m} \left(\frac{y_a}{1+\alpha} - Y_a \right) \qquad \text{(C-8)}$$

Since corrosion is based on the concentration, which can be very different from the mass fraction, especially if the volume of the chamber changes, the change in concentration is given by

$$\frac{dC}{dt} = \frac{d}{dt} \left(\frac{mY_a}{V} \right) = \frac{\frac{d(mY_a)}{dt} V - m \frac{dV}{dt}}{V^2} \qquad \text{(C-9)}$$

Combining and simplifying gives

$$\frac{dC}{dt} = \frac{(1+\alpha)\dot{m}_b}{V} \left(\frac{y_a}{1+\alpha} - C \frac{\hat{R}}{P_0} T_{in} \right) \qquad \text{(C-10)}$$

Gandhi nondimensionalizes the equations using the initial conditions

$$m = m^* m_0 \qquad \text{(C-11)}$$

C-2

$$V = V^* V_0 \tag{C-12}$$

$$T = T^* T_0 \tag{C-13}$$

$$t = \tau t_b \tag{C-14}$$

$$\dot{m}_b = \dot{m}_b^* \frac{m_0}{t_b} \tag{C-15}$$

$$\Gamma_f^* = (1+\alpha)\frac{\dot{m}_b t_b}{m_0} = (1+\alpha)m_b^* \tag{C-16}$$

to result in the following equations for the ASTM E05 gas collection system

$$\frac{dm^*}{d\tau} = \Gamma_f^* \tag{C-17}$$

$$\frac{dV^*}{d\tau} = T_{in}^* \Gamma_f^* \tag{C-18}$$

$$\frac{dT^*}{d\tau} = \frac{T_{in}^* - T^*}{m^*} \Gamma_f^* \tag{C-19}$$

$$\frac{dY_a}{d\tau} = \frac{\dfrac{y_a}{1+\alpha} - Y_a}{m^*} \Gamma_f^* \tag{C-20}$$

$$\frac{dC_a^*}{d\tau} = \frac{\dfrac{y_a}{1+\alpha} - C_a^*}{m^*} \Gamma_f^* \tag{C-21}$$

According to the Annex of ASTM D 5485, the exposure chamber of the cone corrosimeter has a time constant of approximately 150 s. Another way of thinking about this is that with a specified flow of 4.5 L/min or 0.075 L/s if the gas in the exposure chamber (total volume of 11.2 L) is completely mixed then 0.075/11.2 = 0.007 of the effluent mass in the chamber is removed every second. With this, we can create a very simplified model of the amount of the effluent in the exposure chamber and what amount of the effluent came from which period of the combustion.

We will assume that mixing is instantaneous and that there is no deposition. Obviously, there will be deposition but it is likely not to be a large fraction of the total effluent in the chamber. Using these assumptions, the amount of fire effluent in the chamber is

$$\dot{m}_c = \dot{m}_f - .007m_c \tag{C-22}$$

Where m_c is the mass of fire effluent in the exposure chamber and m_f is the mass coming from the fire.

The dependence of the concentration at any time t_i during the burning phase of the test means that unlike the CNET test in ISO 11907-02, the final concentration is very dependent on the shape of the MLR during the test. Three different possible idealized mass loss curves are shown in Figure C-1. Each burns for 600 s and has a peak HRR of 2 MW/m^2. The fraction of the mass flowing into the chamber that is product a is shown in Figure C-1.

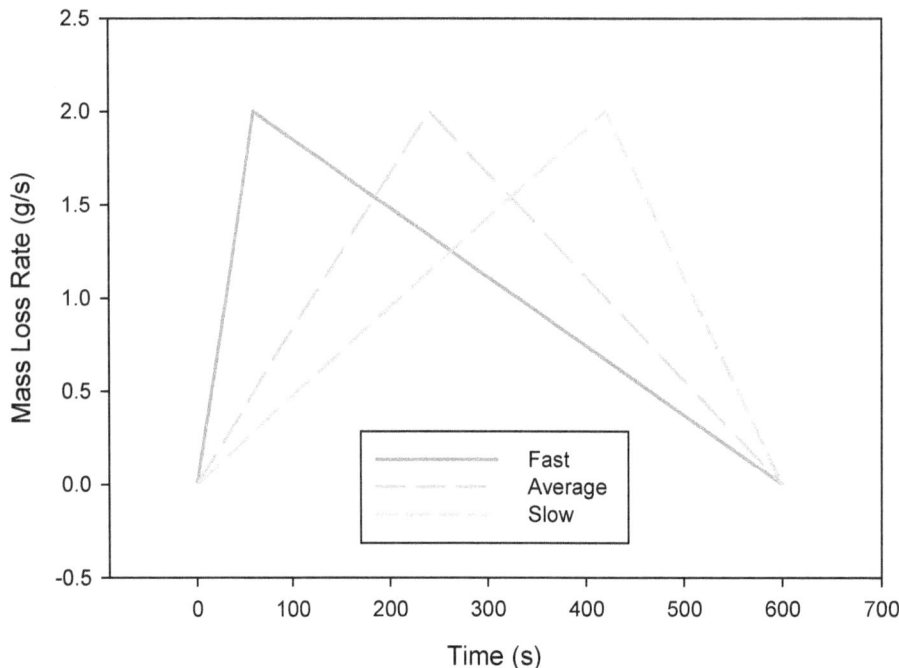

Figure C-1. Three possible mass loss curves as examples of idealized fires in an arbitrary corrosion test scenario.

All three idealized fires lose a total of 600 g in 600 s during combustion. The difference is the time of the peak mass loss rate, which is 60 s for the fast fire, 240 s for the average fire and 420 s for the slow fire. Keeping with the standard that the exposure chamber is closed (the input and output from the exposure chamber closes) when 70 % of the mass has been lost in the fire gives shut down times of 288 s for the Fast fire, 345 s for the Average fire and 420 s for the Slow fire.

Figure C-2 integrates the curves in Figure C-1 to yield the total mass in the exposure chamber versus time for the first 10 min the three fires.

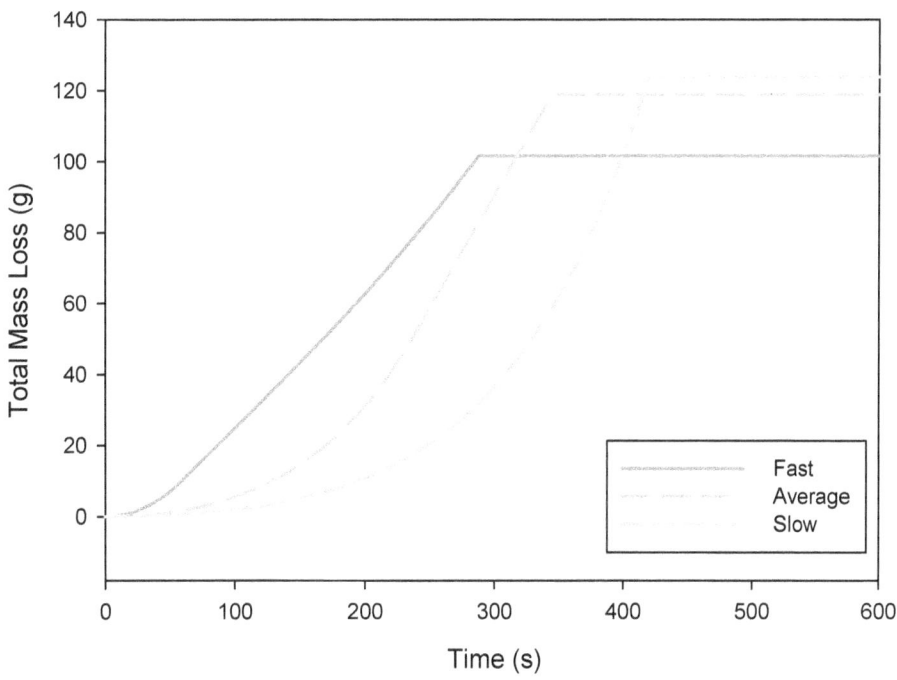

Figure C-2. Total mass in exposure chamber for three idealized fires.

Notice that the fast fire, which is shut down first, has the least mass in the exposure chamber while the slow fire has the most. To answer this question we take advantage of the simplicity of the model to note that the amount of mass that entered the exposure chamber at a particular time t_i is dependent only on the difference between that time and the current time t_c. This means we can determine how much of the total mass in the exposure chamber at the point where the chamber is closed off with the following equation

$$m_c = \dot{m}_f e^{a(t_c - t_i)} \tag{C-23}$$

Where the function $m_c()$ returns the mass in the exposure chamber at the point the total mass lost reaches 70 % of the total mass that will be lost and t_c is the time when the 70 % point is reached and $a = -0.0067$ gives the instantaneous rate that mass is being lost from the exposure chamber. The mass left in the exposure chamber for these three fires is given in Figure C-3.

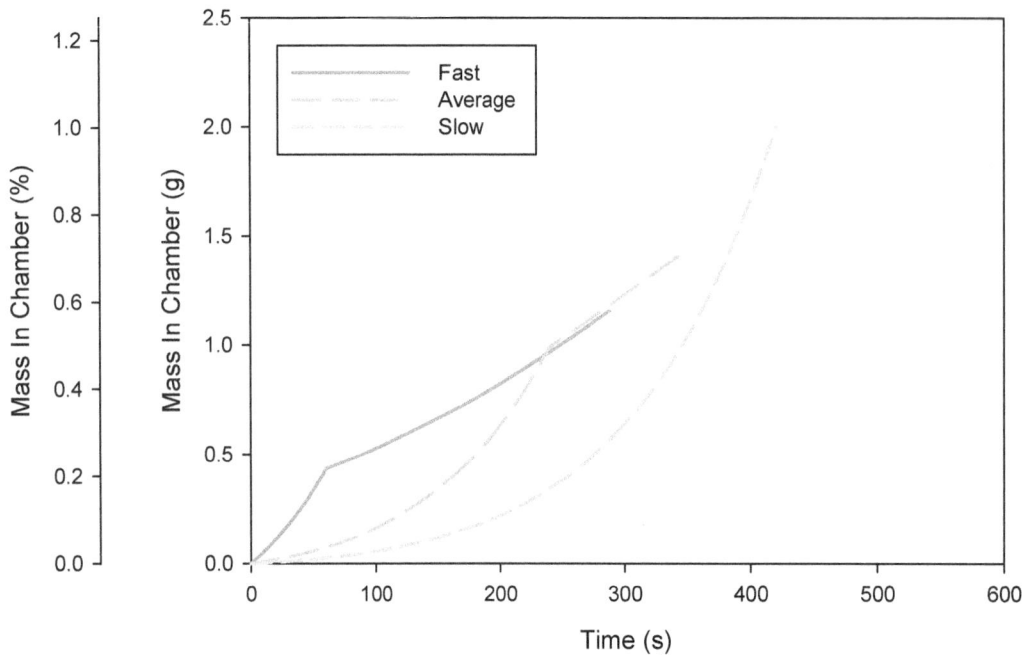

Figure C-3. Mass left in exposure chamber expressed as total mass and as a percentage of mass loss rate for three idealized fires.

For all three fires the peak rate that mass is entering the chamber is 2 g/s. Less than a quarter of that mass is still in the chamber for the fast fire, half the mass is still in the chamber for the average fire but all of the mass is still in the chamber for the slow fire that is closed at the peak rate. Because so much time between when the peak amount of mass enters the chamber before the chamber closes in the fast fire, 228 s, there is more time for that mass to escape. In the slow fire, the peak amount enters the chamber and none of it is allowed to escape. Figure C-3 also shows the percentage of the total mass left in the exposure chamber is from each time period.

One last question is would the shape of the MLR curve effect which material would be the most corrosive in the cone corrosimeter. Corrosion is a very dependent on local conditions and so is difficult to model on a large scale. However Tewarson [20] found that in his real scale tests corrosion scaled with the 0.6 power of concentration. So using that model and making the assumption that all the mass has the same level of corrosivity we can get an idea of the impact of shape on the total corrosivity. Figure C-4 shows the corrosivity calculated for the three fire sources. For the first 10 mins (inset), there is a noticeable difference in the three curves with the fast fire being the most corrosive and the slow fire the least. In the full hour of testing, the order changes. At the end of an hour the medium and slow fires are nominally identical and the fast fire is noticeable less corrosive. This suggests that shape of the mass loss curve can affect the overall corrosiveness of the material being tested.

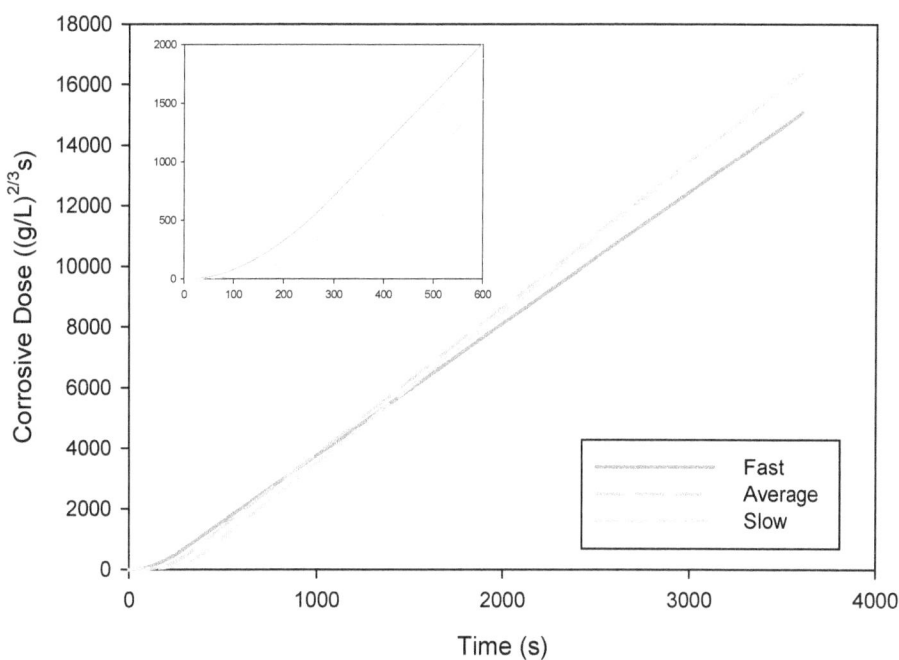

Figure C-4. Estimated corrosive dose for three different idealized fire growth rates.

APPENDIX D CONDUCTIVITY AND CORROSION

Typically, conductivity of combustion products is seen as an indicator of the potential for corrosion. The published data from the PFPC study [72,75-77] allows for more analysis to be performed. With test results from 24 materials it is possible to test if the strong correlation seen by Bottin [71] exists in these data. A plot of the conductivity as measured in DIN 57 472 against the results of CNET testing in Figure D-1 shows the relationship.

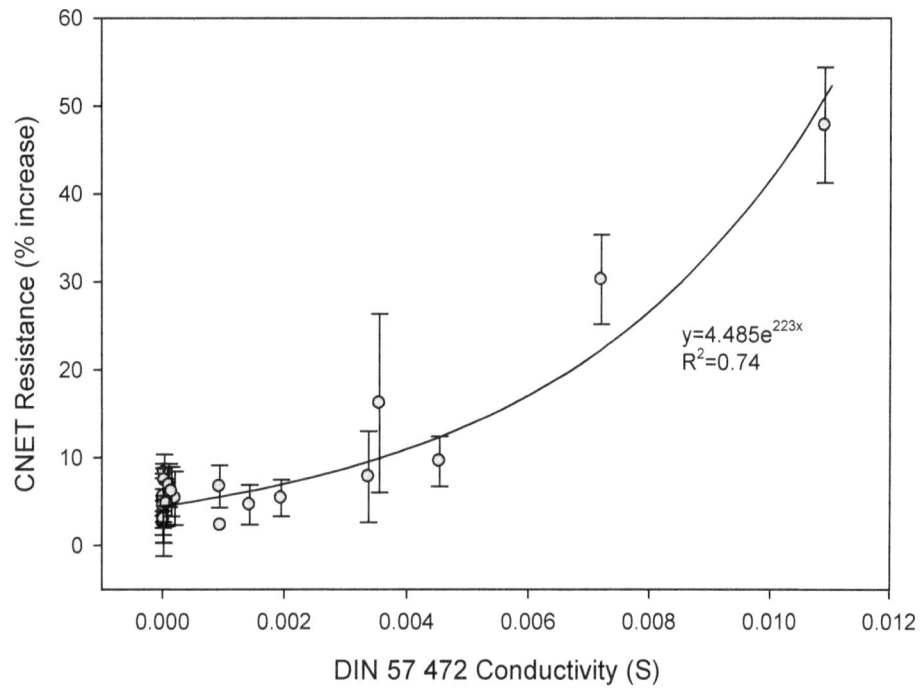

Figure D-1. Comparison of CNET test results with conductivity as measured by DIN 57 472. Data from reference [75,76].

While there is a clear correlation between conductivity and the increase in resistance in the CNET test, the correlation is different from the Bottin data with exponent of approximately 7000 Siemens for the Bottin data being more than 30 times the PFPC exponent of approximately 200 S. Still, the same functional relationship exists in both sets of data.

In the CNET data, there is a group of materials with conductivity less than 250 µS. Focusing on this data shows little correlation but the data does have bounds as can be seen in Figure D-2.

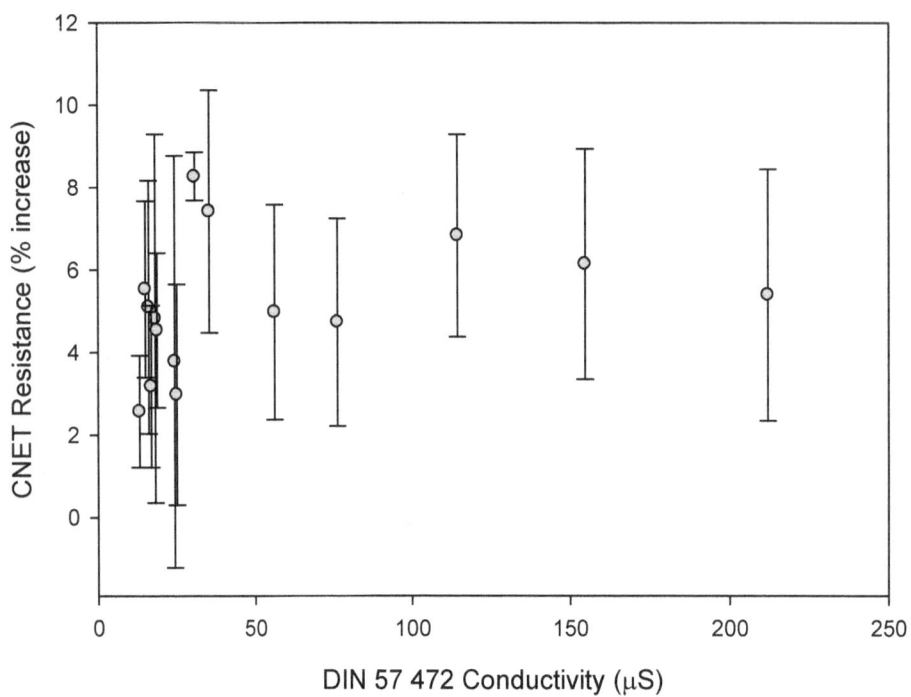

Figure D-2. Comparison of CNET test results with conductivity as measured by DIN 57 472 for materials with conductivity less than 250 µS. Data from reference [75,76].

The resistance change is less than10 % and the conductivity is less than 250 µS. This suggests that when an effluent that has low conductivity when dissolved in water, the material does not enhance the corrosion that occurs from the natural environment of a fire.

Figure D-3 shows a similar comparison for the cone corrosimeter and the E05.21.71 test. For materials with conductivity greater than 900 µS, the figure shows a correlation with the corrosimeter data though not as strong as the CNET data. Because the ASTM E05.21.70 test used a probe that was only 2,500 Å thick, values more corrosive materials all show values near 2,500 Å. This indicates that all the material on the probe was corroded and thus no correlation is expected. The same lack of correlation exists for materials that have a particularly low conductance.

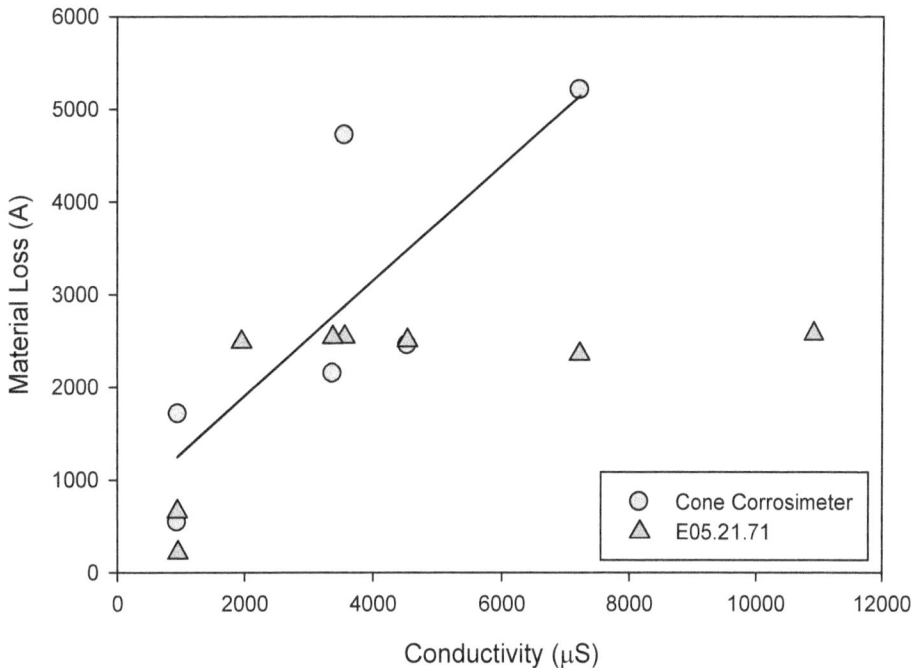

Figure D-3. Comparison of cone corrosimeter and E05.21.71 test results with conductivity as measured by DIN 57 472. Data from reference [76,77].

Finally, we will consider how the CNET test results compare to the cone corrosimeter results. Figure D-4 shows a comparison the corrosimeter data against CNET data. The data breaks up into two groups with the upper group showing a reasonably linear correlation. The lower group is not well correlated but it is bounded and represents minor corrosion in both tests. If we just look at highly conductive materials, all the strongly correlated values of the upper group are included as well as one from the lower group. It is interesting to note that the *pH* value for this product, product 5.6, the highest *pH* of any of the high conductivity group. This correlation has been seen before by Gandhi [100] in his review of the data from the PFPC study, suggesting that *pH* has some role in predicting corrosion.

To look at the relationship between *pH* and conductivity, it is reasonable to look at the same transform on the conductivity as is done to get the *pH*

$$pS = -\log\left(\frac{c}{10^6}\right) \tag{D-1}$$

Figure D-5 shows the data with the high conductance and low conductance materials plotted as two different groups.

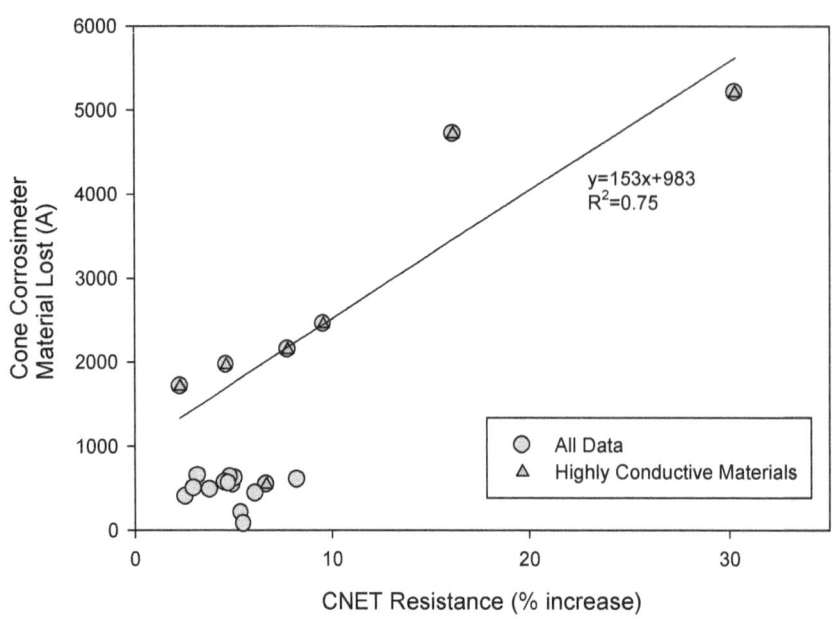

Figure D-4. Comparison of CNET test and cone corrosimeter test results for several materials. Data from reference [75,77].

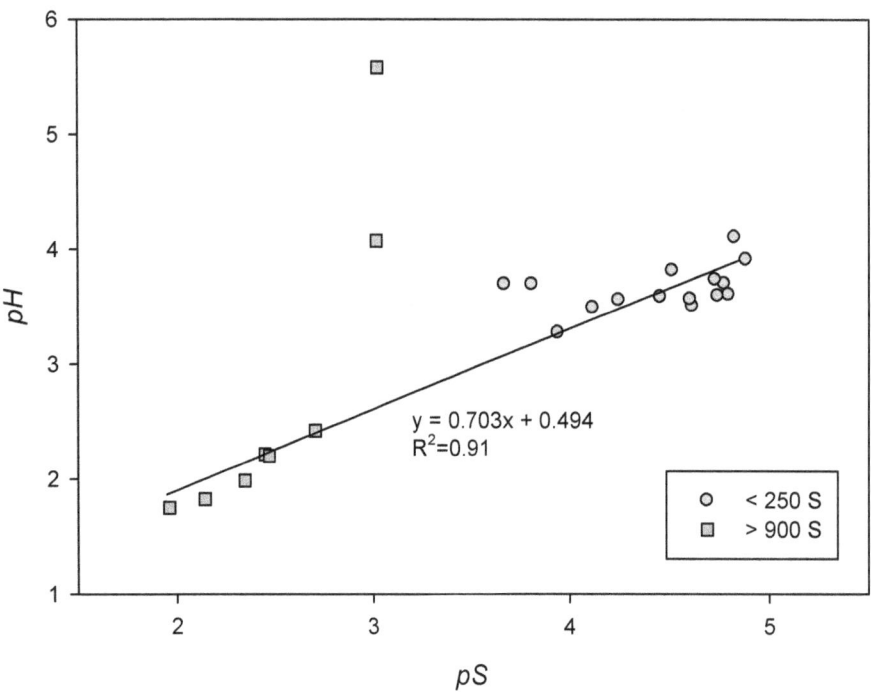

Figure D-5. Comparison of acidity and conductivity for several materials from the PFPC Study. Data from reference [76].

It is important to remember that the lower the number the higher the acidity or conductance. There is a trend in the data that linearly relates *pH* and *pS* with clearly some outlying data. Several of the polymer products have a component that is *pH* neutral but conductive given a relatively high pH but a lower conductance. So there is not necessarily a relationship between conductance and *pH*. Conductance being the driving force also answers the question of how bases can also be very corrosive.

Even with 24 materials this is but a small sampling of possible materials. While the relationship between conductance and corrosion seems to be a strong one, it still warrants additional research.

NRC FORM 335 (12-2010) NRCMD 3.7	U.S. NUCLEAR REGULATORY COMMISSION	1. REPORT NUMBER (Assigned by NRC, Add Vol., Supp., Rev., and Addendum Numbers, if any.)
BIBLIOGRAPHIC DATA SHEET *(See instructions on the reverse)*		NUREG/CR-7123

2. TITLE AND SUBTITLE	3. DATE REPORT PUBLISHED	
A Literature Review of the Effects of Smoke from a Fire on Electrical Equipment	MONTH	YEAR
	July	2012
	4. FIN OR GRANT NUMBER	
	N6761	

5. AUTHOR(S)	6. TYPE OF REPORT
R.D. Peacock, T.G. Cleary, P.A. Reneke, and D.C. Murphy	Technical
	7. PERIOD COVERED (Inclusive Dates)
	9/9/2009 - 7/30/2011

8. PERFORMING ORGANIZATION - NAME AND ADDRESS (If NRC, provide Division, Office or Region, U. S. Nuclear Regulatory Commission, and mailing address; if contractor, provide name and mailing address.)

National Institute of Standards and Technology
Engineering Laboratory
100 Bureau Drive, Stop 8664
Gaithersburg, MD 20899-8664

9. SPONSORING ORGANIZATION - NAME AND ADDRESS (If NRC, type "Same as above", if contractor, provide NRC Division, Office or Region, U. S. Nuclear Regulatory Commission, and mailing address.)

Division of Risk Analysis
Office of Nuclear Regulatory Research
U.S. Nuclear Regulatory Commission
Washington, DC 20555-0001

10. SUPPLEMENTARY NOTES

D.W. Stroup, NRC Project Manager

11. ABSTRACT (200 words or less)

A review is presented of the state of the art of smoke production measurement, prediction of smoke impact as part of computer-based fire modeling, and measurement and prediction of the impact of smoke through deposition of soot on and corrosion of electrical equipment. The literature review on smoke corrosivity testing and damage due to smoke deposition emphasizes (despite extensive research on smoke corrosivity) the lack of validated and widely applicable prescriptive or performance based methods to assure electrical equipment survivability given exposure to smoke from a fire. Circuit bridging via current leakage through deposited smoke was identified as a potentially important mechanism of electronic and electrical equipment failure during nuclear power plant fires.

In the near term, assessment of potential damage can reasonably be based on the airborne smoke exposure concentration and the exposure duration. Hence, models that can predict the airborne smoke concentration would be sufficient to provide upper limit estimates of potential damage. In the longer term, it would be desirable to develop models that could estimate the deposition behavior of smoke, and specifically correlate the combination of deposited and airborne smoke to component damage.

12. KEY WORDS/DESCRIPTORS (List words or phrases that will assist researchers in locating the report.)	13. AVAILABILITY STATEMENT
corrosivity, electrical equipment, equipment damage, literature review, smoke movement, smoke transport	unlimited
	14. SECURITY CLASSIFICATION
	(This Page) unclassified
	(This Report) unclassified
	15. NUMBER OF PAGES
	16. PRICE

NRC FORM 335 (12-2010)

UNITED STATES
NUCLEAR REGULATORY COMMISSION
WASHINGTON, DC 20555-0001

_____

OFFICIAL BUSINESS

NUREG/CR-7123

A Literature Review of the Effects of Smoke from a Fire on Electrical Equipment

July 2012